GENETICS – RESEARCH AND ISSUES

DNA

BACKGROUND, LAWS AND BACKLOG OF EVIDENCE

GENETICS – RESEARCH AND ISSUES

Additional books and e-books in this series can be found on Nova's website under the Series tab.

GENETICS – RESEARCH AND ISSUES

DNA

BACKGROUND, LAWS AND BACKLOG OF EVIDENCE

TOMÁŠ KOLÁČEK
EDITOR

snova
New York

Library of Congress Cataloging-in-Publication Data

ISBN: 978-1-53616-117-5

Published by Nova Science Publishers, Inc. † New York

CONTENTS

PREFACE

Deoxyribonucleic acid, or DNA, is the fundamental building block for an individual's entire genetic makeup. DNA is a powerful tool for law enforcement investigations because each person's DNA is different from that of every other individual (except for identical twins). As early as the 1980s, states began enacting laws that required the collection of DNA samples from offenders convicted of certain sexual and other violent crimes. Chapter 1 provides an overview of how DNA is used to investigate crimes and help protect the innocent. Chapters 2 and 3 report on the establishment of a system for integration of Rapid DNA instruments for use by law enforcement to reduce violent crime and reduce the current DNA analysis backlog. Chapter 4 examines what is known about the amount of backlogged DNA evidence at state and local government labs; the extent to which OJP measures CEBR grant performance; and the extent to which OJP has designed controls to identify conflicts of interest related to CEBR grants. Chapter 5 reviews the level of crime scene DNA evidence backlogs among CEBR grantees and the factors that contribute to such backlogs; the extent to which DOJ has clearly defined goals for CEBR; and the extent to which OJP has controls for CEBR related to federal conflicts of interest and lobbying requirements. In 2016, about 323,000 individuals age 12 or older were reported victims of sexual assault, according to the Bureau of Justice Statistics. Studies have shown that exams performed by sexual assault forensic examiners—medical providers trained in collecting and preserving forensic evidence—may result in better physical and mental health care for victims, better evidence collection, and higher prosecution rates. Chapter 6 describes what was known in 2016 about the availability of sexual assault forensic examiners nationally and in selected states and the challenges selected states faced in maintaining a supply of sexual assault forensic examiners.

Chapter 1 - Deoxyribonucleic acid, or DNA, is the fundamental building block for an individual's entire genetic makeup. DNA is a powerful tool for law enforcement investigations because each person's DNA is different from that of every other individual (except for identical twins). DNA can be extracted from a number of sources, such as hair, bone, teeth, saliva, and blood. As early as the 1980s, states began enacting laws that required the collection of DNA samples from offenders convicted of certain sexual and other violent crimes. The samples are analyzed and their profiles entered into state databases. In the late 1980s, the Federal Bureau of Investigation (FBI) Laboratory convened a working group of federal, state, and local forensic scientists to establish guidelines for the use of forensic DNA analysis in laboratories. The group proposed guidelines that are the basis of current national quality assurance standards, and it urged the creation of a national DNA database. The criminal justice community began to utilize DNA analyses more often in criminal investigations and trials, and in 1994, Congress enacted legislation to authorize the creation of a national DNA database. Federal law (34 U.S.C §12592(a)) authorizes the FBI to operate and maintain a national DNA database where DNA profiles generated from samples collected from people under applicable legal authority and samples collected at crime scenes can be compared to generate leads in criminal investigations. Statutory provisions also authorize the collection of DNA samples from federal offenders and arrestees, District of Columbia offenders, and military offenders. State laws dictate which convicted offenders, and in some states arrestees, will have profiles entered into state DNA databases, while federal law dictates the scope of the national database. Increasing awareness of the power of DNA to solve crimes has resulted in increased demand for DNA analysis, which has resulted in a backlog of casework.

Some jurisdictions have started to use their DNA databases for familial searching, which involves using offender profiles to identify relatives who might be perpetrators of crimes. In addition to solving crimes, DNA analysis can also help exonerate people incarcerated for crimes they did not commit. Congress has authorized several grant programs to provide assistance to state and local governments for forensic sciences. Many of the programs focus on providing state and local governments with funding to reduce the backlog of forensic and convicted offender DNA samples waiting to be processed and entered into the national database. Other grant programs provide funding for related purposes, such as offsetting the cost of providing postconviction DNA testing.

Chapter 2 - This is an edited, reformatted and augmented version of House of Representatives, 115th Congress, 1st Session, Publication No. Report 115–117, dated May 11, 2017.

Chapter 3 - This is an edited, reformatted and augmented version of 115th Congress, 2^{d} Session, House of Representatives, Publication No. Report 115–504, dated January 9, 2018.

Chapter 4 - Many state and local government crime labs continue to have backlogs of requests for crime scene DNA analysis, which may include sexual assault kits, despite DOJ awarding nearly $1 billion since 2004 through the CEBR grant program to enhance lab capacity and reduce backlogs. Additionally, questions have been raised about potential improper connections among those who profit from CEBR grants—such as private labs and DNA equipment vendors— and those who advocate for CEBR funding. This chapter examines, among other things, (1) what is known about the amount of backlogged DNA evidence at state and local government labs; (2) the extent to which OJP measures CEBR grant performance; and (3) the extent to which OJP has designed controls to identify conflicts of interest related to CEBR grants. GAO reviewed CEBR grantee data on DNA evidence backlogs from 2011-2017 (the most recent years of comparable data). GAO reviewed documentation on CEBR performance and controls related to conflicts of interest, transparency in grantee procurement, and lobbying. GAO also reviewed relevant reports and studies, and interviewed officials from DOJ.

Chapter 5 - Many state and local crime labs have backlogs of requests for DNA analysis of crime scene evidence, as reported by grantees participating in DOJ's CEBR grant program. These backlogs can include sexual assault kits. Since 2011, DOJ's Office of Justice Programs—the primary grant-making arm of DOJ—has awarded nearly $500 million to states and local jurisdictions through the CEBR grant program to help reduce DNA evidence awaiting analysis at crime labs. There have been concerns that these backlogs of unanalyzed evidence have enabled serial offenders to reoffend or have delayed justice. This statement is based on preliminary observations and analyses from GAO's ongoing review of (1) the level of crime scene DNA evidence backlogs among CEBR grantees and the factors that contribute to such backlogs; (2) the extent to which DOJ has clearly defined goals for CEBR; and (3) the extent to which OJP has controls for CEBR related to federal conflicts of interest and lobbying requirements. To develop these preliminary findings, GAO reviewed CEBR grantee data from 2011-2016 (the latest data available) and studies relevant to the DNA backlog, visited selected labs, and interviewed DOJ officials, among others.

Chapter 6 - In 2016, about 323,000 individuals age 12 or older were reported victims of sexual assault, according to the Bureau of Justice Statistics. Studies have shown that exams performed by sexual assault forensic examiners—medical providers trained in collecting and preserving forensic evidence—may result in better physical and mental health care for victims, better evidence collection, and higher prosecution rates. Yet, concerns have been raised about the availability of such examiners. The Department of Justice administers grant programs that can be used by states and other eligible entities to train and fund examiners. This statement summarizes GAO's findings from its March 2016 report (GAO-16-334) describing (1) what was known in 2016 about the availability of sexual assault forensic examiners nationally and in selected states and (2) the challenges selected states faced in maintaining a supply of sexual assault forensic examiners. For that report, GAO reviewed literature on the availability of examiners and challenges training and retaining them. GAO also interviewed knowledgeable officials, including recipients of federal sexual assault examiner related grants and officials from sexual assault coalitions in six states (Colorado, Florida, Massachusetts, Nebraska, Oregon, and Wisconsin) selected to achieve variation in factors such as population and geographic location.

In: DNA
Editor: Tomáš Koláček
ISBN: 978-1-53616-117-5

Chapter 1

DNA TESTING IN CRIMINAL JUSTICE: BACKGROUND, CURRENT LAW, AND GRANTS*

Nathan James

SUMMARY

Deoxyribonucleic acid, or DNA, is the fundamental building block for an individual's entire genetic makeup. DNA is a powerful tool for law enforcement investigations because each person's DNA is different from that of every other individual (except for identical twins). DNA can be extracted from a number of sources, such as hair, bone, teeth, saliva, and blood. As early as the 1980s, states began enacting laws that required the collection of DNA samples from offenders convicted of certain sexual and other violent crimes. The samples are analyzed and their profiles entered into state databases. In the late 1980s, the Federal Bureau of Investigation (FBI) Laboratory convened a working group of federal, state, and local forensic scientists to establish guidelines for the use of forensic DNA analysis in laboratories. The group proposed guidelines that are the basis of current national quality assurance standards, and it

* This is an edited, reformatted and augmented version of a Congressional Research Service publication R41800, prepared for Members and Committees of Congress dated January 24, 2018.

urged the creation of a national DNA database. The criminal justice community began to utilize DNA analyses more often in criminal investigations and trials, and in 1994, Congress enacted legislation to authorize the creation of a national DNA database.

Federal law (34 U.S.C §12592(a)) authorizes the FBI to operate and maintain a national DNA database where DNA profiles generated from samples collected from people under applicable legal authority and samples collected at crime scenes can be compared to generate leads in criminal investigations. Statutory provisions also authorize the collection of DNA samples from federal offenders and arrestees, District of Columbia offenders, and military offenders. State laws dictate which convicted offenders, and in some states arrestees, will have profiles entered into state DNA databases, while federal law dictates the scope of the national database. Increasing awareness of the power of DNA to solve crimes has resulted in increased demand for DNA analysis, which has resulted in a backlog of casework. Some jurisdictions have started to use their DNA databases for familial searching, which involves using offender profiles to identify relatives who might be perpetrators of crimes. In addition to solving crimes, DNA analysis can also help exonerate people incarcerated for crimes they did not commit.

Congress has authorized several grant programs to provide assistance to state and local governments for forensic sciences. Many of the programs focus on providing state and local governments with funding to reduce the backlog of forensic and convicted offender DNA samples waiting to be processed and entered into the national database. Other grant programs provide funding for related purposes, such as offsetting the cost of providing postconviction DNA testing.

INTRODUCTION

Deoxyribonucleic acid, or DNA, is the fundamental building block for an individual's entire genetic makeup. DNA is a powerful tool for law enforcement investigations because each person's DNA is different from that of every other individual (except for identical twins). By analyzing selected DNA sequences (called loci), a crime laboratory can develop a profile to be used in identifying a suspect.

DNA can be extracted from a number of sources, such as hair, bone, teeth, saliva, and blood. Because the human body contains so many copies of DNA, even a minuscule amount of bodily fluid or tissue can yield useful information. Obtaining a DNA sample is not complicated; it can be as simple as a swab of the inside of the mouth to obtain cheek cells and white blood cells in saliva.

State and federal DNA databases have proved instrumental in solving crimes, reducing the risk of convicting the wrong person, and establishing the innocence of those wrongly convicted. DNA evidence is used to solve crimes in two ways:

- In cases where a suspect is known, a sample of that person's DNA can be compared to biological evidence found at a crime scene. The results of this comparison may then help establish whether the suspect was at the crime scene or whether he/she committed the crime.
- In cases where a suspect is not known, biological evidence from the crime scene can be analyzed and compared to offender profiles contained in existing DNA databases to assist in identifying the perpetrator. Through the use of DNA databases, biological evidence found at one crime scene can also be connected to other crime scenes, linking them to the same perpetrator or perpetrators.

This chapter provides an overview of how DNA is used to investigate crimes and help protect the innocent.[1] It also reviews current statutory law on collecting DNA samples, sharing DNA profiles generated from those samples, and providing access to postconviction DNA testing. The report also includes a summary of grant programs authorized by Congress to assist state and local governments with reducing DNA backlogs, provide postconviction DNA testing, and promote new technology in the field.

---

[1] This report does not include a discussion of the use of DNA to identify missing persons and unidentified human remains, nor does it include an overview of grant programs to state and local governments for developing DNA profiles from samples from missing persons, close relatives of missing persons, or unidentified human remains. For more on this issue, see CRS Report RL34616, *Missing Adults: Background, Federal Programs, and Issues for Congress.*

BACKGROUND

Federal law authorizes the Federal Bureau of Investigation (FBI) to operate and maintain a national DNA database where DNA profiles generated from samples collected from people under applicable legal authority and samples collected at crime scenes can be compared to generate leads in criminal investigations. Statutory provisions also authorize the collection of DNA samples from federal offenders and arrestees, District of Columbia offenders, and military offenders. State law dictates which arrestees and convicted offenders will have profiles entered into state DNA databases, but federal law dictates which profiles entered into state databases can be uploaded into the national DNA database.

Increased awareness of the power of DNA testing to solve crimes has led to increased demand for DNA analysis, which has resulted in a backlog of casework for laboratory personnel. In addition to solving crimes, DNA analysis can also help exonerate people incarcerated for crimes they did not commit.

The National DNA Index System (NDIS) and the Combined DNA Index System (CODIS)

As early as the 1980s, states began enacting laws that required DNA samples from offenders convicted of certain sexual offenses and other violent crimes. The samples were then analyzed and their profiles entered into state databases. In the late 1980s, the FBI Laboratory convened a working group of federal, state, and local forensic scientists to establish guidelines for the use of forensic DNA analysis in laboratories. The group proposed guidelines that are the basis of current national quality assurance standards, and it urged the creation of a national DNA database.[2] In 1994,

[2] Statement of Dwight E. Adams, Deputy Assistant Director, Laboratory Division, Federal Bureau of Investigation, in U.S. Congress, House of Representatives, Government Reform Committee, Subcommittee on Government Efficiency, Financial Management and Intergovernmental Relations, *How Effective are State and Federal Agencies Working Together to Implement the Use of New DNA Technologies*?, hearing, 107th Cong., 1st sess., March 29, 2004, pp. 53-54.

Congress authorized the FBI to establish and oversee the National DNA Index System (NDIS). When the NDIS launched in 1998, nine states participated.[3] Currently, laboratories in all 50 states, the District of Columbia, the federal government, Puerto Rico, and the U.S. Army Criminal Investigation Laboratory participate in the NDIS.[4] The NDIS contains the DNA profiles provided by federal, state, and participating local crime laboratories.[5]

DNA profiles generated by laboratories operated by local law enforcement agencies are stored in Local DNA Index Systems (LDIS). DNA profiles generated by state laboratories, along with authorized profiles stored in participating LDIS, are uploaded into State DNA Index Systems (SDIS). Each state has its own laws specifying which profiles can be included in the SDIS.[6] DNA profiles generated by federal laboratories, along with authorized DNA profiles in participating SDIS, are uploaded into the NDIS.[7] Federal law dictates which DNA profiles can be stored in the NDIS (see below). The NDIS allows participating laboratories to compare DNA on the national level while the SDIS allows each state to compare DNA profiles stored at the state level. Federal, state, and local laboratories upload and compare DNA profiles using the Combined DNA Index System (CODIS) software produced and distributed by the FBI.[8]

CODIS searches three indexes (convicted offenders, arrestee, and forensic) to generate investigative leads. The convicted offender index

---

[3] John M. Butler, *Fundamentals of Forensic DNA Typing* (Burlington, MA: Academic Press, 2010), p. 265 (hereinafter, *Fundamentals of Forensic DNA Typing*).

[4] Information on the number of participant laboratories and the number of profiles they have uploaded into the NDIS can be found on the FBI's website at https://www.fbi.gov/services/laboratory/biometric-analysis/codis/ndis-statistics.

[5] U.S. Department of Justice, Federal Bureau of Investigation, *Frequently Asked Questions (FAQs) on the CODIS Program and the National DNA Index System*, https://www.fbi.gov/services/laboratory/biometric-analysis/codis/codis- and-ndis-fact-sheet (hereinafter "CODIS FAQs").

[6] The National Conference of State Legislatures (NCSL) maintains a searchable database of state DNA laws, including laws related to which convicted offenders are required to submit a sample for inclusion in the state's DNA database and whether, and if so, from whom, collects DNA samples from individuals arrested for certain crimes. The NCSL's database is available online at http://www.ncsl.org/research/civil-and-criminal-justice/dna-laws-database.aspx.

[7] U.S. Department of Justice, Federal Bureau of Investigation, *CODIS—NDIS Statistics*, https://www.fbi.gov/services/laboratory/biometric-analysis/codis/ndis-statistics.

[8] CODIS FAQs.

contains DNA profiles developed from samples collected from convicted offenders; the arrestee index contains DNA profiles developed from samples collected from arrested but *not yet convicted* individuals; and the forensic index contains DNA profiles developed from samples collected at crime scenes. CODIS searches across these indexes to look for potential matches (also referred to as "hits").[9] Matches can occur between either the convicted offender or arrestee indexes and the forensic index, thereby providing law enforcement with the identity of one or more suspects.[10] Also, matches can occur between DNA profiles in the forensic index, thereby linking crime scenes to each other and identifying serial offenders.[11] Matches between multiple samples in the forensic index can allow law enforcement agencies in different jurisdictions to coordinate their efforts and share leads. No names or other personal identifiers for offender and arrestee DNA profiles are stored in the NDIS, so when a match is made in CODIS, the laboratories that submitted the DNA profiles to the NDIS are notified of the match and they contact each other to verify the match and coordinate their efforts.[12]

DNA Profiles

DNA profiles entered into CODIS are based on 13 core short tandem repeat (STR) loci selected by the FBI.[13] Currently, the 13 STR loci used by the FBI are non-coding, meaning that they have not been shown to be associated with human attributes such as height, eye or skin color, or susceptibility to a particular disease.[14] Each locus has two alleles, and it is these 13 pairs of alleles that are compared to match samples in the forensic

---

[9] Ibid.

[10] Ibid. If an "offender hit" is obtained, that information typically is used as probable cause to obtain a new DNA sample from that suspect so the match can be confirmed by the crime laboratory before an arrest is made.

[11] Ibid.

[12] CODIS FAQs.

[13] Ibid.

[14] Jules Epstein, "Genetic Surveillance—The *Bogeyman* Response to Familial DNA Investigations," *University of Illinois Journal of Law, Technology and Policy*, vol. 2009, no. 1, (2009), p. 143.

index with profiles in either the offender or arrestee indexes. The 13 core loci chosen by the FBI provide a high level of discriminatory power. The probability that two unrelated individuals would share all 13 pairs of alleles is estimated to be one in several hundred billion.[15] Two random Americans will, on average, share two or three alleles.[16]

It is important to ensure the quality of the DNA profiles entered into the NDIS. If the profiles are not accurate, they are of little use for making matches between forensic and offender or arrestee profiles. The FBI helps ensure the quality of DNA profiles included in the NDIS by signing memorandums of understanding with state laboratories whereby the laboratory agrees to adhere to the FBI's Quality Assurance Standards (QAS, see below).[17] Laboratories submitting DNA profiles to the NDIS must be accredited and audited annually.[18] Annual audits can be conducted by either an internal or external auditor, but laboratories must be audited by an external agency at least once every two years.[19] Laboratories that do not pass the annual audit can be prevented from entering DNA profiles in CODIS.[20] Currently, most labs in the United States are to be audited by the American Society of Crime Laboratory Directors and its Laboratory Accreditation Board (ASCLD/LAB) and Forensic Quality Services (FQS). In addition, DNA analysts must undergo semiannual proficiency testing.[21] DNA analysts who do not pass their semiannual proficiency tests are not to be allowed to enter profiles into CODIS.[22] Laboratories are also required to conduct two reviews of all DNA profiles before they are entered into

---

[15] Henry T. Greely, Daniel P. Riordan, and Nanibaa' A. Garrison et al., "Family Ties: The Use of DNA Offender Databases to Catch Offenders' Kin," *Journal of Law, Medicine and Ethics*, vol. 34, no. 2 (Summer 2006), p. 250 (hereinafter, "Greely, Riordan, Garrison et al., 'Family Ties'").

[16] Ibid.

[17] *Fundamentals of Forensic DNA Typing*, p. 270.

[18] Ibid., p. 271.

[19] U.S. Department of Justice, Federal Bureau of Investigation, *Quality Assurance Standards for DNA Databasing Laboratories*, Standard 15, https://www.fbi.gov/file-repository/ quality-assurance-standards-for-dna-databasing- laboratories.pdf/view. U.S. Department of Justice, Federal Bureau of Investigation, *Quality Assurance Standards for Forensic DNA Testing Laboratories*, Standard 15, https://www.fbi.gov/file-repository/quality-assurance-standards-for- forensic-dna-testing-laboratories.pdf/view (hereinafter "QAS").

[20] *undamentals of Forensic DNA Typing*, p. 271.

[21] Ibid.

[22] Ibid.

CODIS.[23] Currently, as prescribed by federal law (see below), only public laboratories that comply with the QAS can submit DNA profiles to the NDIS. However, public laboratories are allowed to outsource casework to private laboratories. All private laboratories that conduct DNA testing for public laboratories must be accredited, be audited annually, and adhere to the requirements of the QAS.[24] Public laboratories are required to conduct an initial site visit to each private laboratory they contract with to conduct DNA analyses.[25] If the public laboratory signs a contract with a private laboratory that is longer than one year, the public laboratory must conduct an annual site visit.[26] Public laboratories are also required to review all outsourced DNA profiles generated by private laboratories.[27] The review by the public laboratory is in addition to the two reviews private laboratories are required to conduct per the QAS.

An offender or arrestee profile in a DNA database consists of 26 numbers representing each of the two alleles for the 13 STR loci, an agency identification number, a sample identification number, and an identifier for the analyst that entered the information.[28] However, most jurisdictions retain the DNA sample used to generate the profile placed in CODIS.[29] DNA samples are usually retained for quality assurance purposes, such as confirming a hit made using the NDIS, and it allows jurisdictions to retest the sample if new technology is developed in the future.[30] Privacy advocates are concerned that stored DNA samples include a wealth of genetic information that could be misused.[31] States and the federal government have sought to prevent the unauthorized use of DNA

---

[23] U.S. Congress, House Committee on the Judiciary, Subcommittee on Crime, Terrorism, and Homeland Security, *Testimony of Jeffery S. Boschwitz, Ph.D.*, Hearing on "Rape Kit Backlogs: Failing the Test of Providing Justice to Sexual Assault Survivors", 111th Cong., 2nd sess., May 20, 2010, H.Hrg 111-115 (Washington: GPO, 2010), p. 81.

[24] QAS, Standard 17.

[25] CODIS FAQs.

[26] QAS, Standard 17.

[27] Ibid.

[28] *Fundamentals of Forensic DNA Typing*, p. 270.

[29] Ibid., p. 262.

[30] Ibid.

[31] Tania Simoncelli, "Dangerous Excursions: The Case Against Expanding Forensic DNA Databases to Innocent Persons," *Journal of Law, Medicine, and Ethics*, vol. 34, no. 2 (Summer 2006), p. 392.

samples. Some states have criminal penalties in place for individuals who misuse DNA samples collected for law enforcement purposes.[32] Under current law, anyone who misuses a DNA sample collected under federal authority is subject to a fine of up to $250,000, or imprisonment for up to one year.[33]

The number of offender profiles included in the NDIS has increased as Congress has allowed states to include DNA profiles from a broader range of convicted offenders and persons arrested for certain crimes to be included in the database. States have also amended their DNA collection laws to reflect this expanded authority. Approximately 15.6 million new convicted offender and arrestee profiles have been added to NDIS since 2000.[34] In addition, approximately 795,000 new forensic profiles have been included in the NDIS since 2000. The additional offender and forensic profiles have increased the number of investigative leads generated by DNA databases. Since the creation of NIDS, hits generated by searches of the NDIS have aided in the investigation of nearly 381,000 crimes.[35]

DNA Backlog

Delays in processing DNA evidence can result in delays in apprehending or prosecuting violent or serial offenders, or they can result in wrongfully convicted individuals serving time in prison for crimes they did not commit. In addition, persistent backlogs can result in crime laboratories prioritizing DNA analysis for violent offenses, such as homicide or sexual assault, over other offenses, such as property crimes, or they can result in law enforcement agencies establishing policies stating

[32] Ibid., p. 392.

[33] 42 U.S.C. §14135e(c).

[34] The FBI reports data on the number of offender, arrestee, and forensic profiles in the NDIS in 2000 at http://www.fbi.gov/about-us/lab/biometric-analysis/codis/codis_brochure. The most recent data on the number of offender, arrestee, and forensics profiles in the NDIS can be found at https://www.fbi.gov/services/laboratory/ biometric-analysis/codis/ndis-statistics. The figures on the number of profiles added since 2000 are based on the reported number of offender, arrestee, and forensic profiles in the NDIS as of November 2017.

[35] U.S. Department of Justice, Federal Bureau of Investigation, *CODIS—NDIS Statistics*, https://www.fbi.gov/services/laboratory/biometric-analysis/codis/ndis-statistics.

that biological evidence is not to be collected for minor offenses.[36] Not analyzing or collecting DNA samples for minor offenses could prevent law enforcement from apprehending offenders who may go on to commit more serious crimes.

Context is important when evaluating data on DNA backlogs.[37] Backlogs are best considered in the context of each crime laboratory's capacity, size, and workload. For example, if there are two laboratories and the first laboratory has a backlog of casework that is three times the size of the casework backlog in the second laboratory, the backlog for the first laboratory might not be as daunting if the first laboratory's turnaround time is twice as fast as the second laboratory and the analysts in the first laboratory are more productive (i.e., each analyst analyzes more cases per month).

Forensic Casework

The most recent data available on the size of DNA backlogs are from a November 2016 report published by the Bureau of Justice Statistics (BJS).[38] Data presented in the report come from BJS's 2014 Census of Publicly Funded Forensic Crime Laboratories (CPFCL). In the report, BJS compares data from the 2014 census to data taken from the 2009 CPFCL to show how requests for DNA analysis have changed.

In 2009, there were an estimated 103,500 backlogged forensic casework analyses, while in 2014 there were an estimated 107,800 backlogged analyses.[39] However, the estimated increase in backlogged

---

[36] Edwin Zedlewski and Mary B. Murphy, "DNA Analysis for 'Minor' Crimes: A Major Benefit for Law Enforcement," *NIJ Journal*, vol. 253 (January 2006) (hereinafter, "DNA Analysis for 'Minor' Crimes").

[37] Mark Nelson, Ruby Chase, and Lindsay DePalma, *Making Sense of DNA Backlog, 2012s—Myths vs. Reality*, U.S. Department of Justice, Office of Justice Programs, National Institute of Justice, NCJ 243347, Washington, DC, December 2013, p. 6.

[38] Matthew R. Durose, Andrea M. Burch, and Kelly Walsh, et al., *Publicly Funded Forensic Crime Laboratories: Resources and Services, 2014*, U.S. Department of Justice, Office of Justice Programs, Bureau of Justice Statistics, NCJ 250151, Washington, DC, November 2016.

[39] Ibid., Table 6. A request for analysis was considered to be backlogged if it was not examined and reported to the submitting agency within 30 days.

forensic casework analyses is not statistically significant.[40] BJS data show that public crime laboratories completed more forensic casework samples in 2014 than 2009. In 2014, public crime laboratories completed an estimated 296,000 requests for forensic casework analysis, up from an estimated 239,000 in 2009.[41] While crime laboratories were able to process more cases, the requests received for forensics casework analysis increased from an estimated 260,000 in 2009 to an estimated 333,000 in 2014.[42]

Convicted Offender and Arrestee Samples

Data from BJS show that there was a significant decrease in the backlog of requests for analysis of convicted offender and arrestee samples. The backlog of these samples decreased from 502,500 in 2009 to 64,800 in 2014.[43] Public crime laboratories processed an estimated 1,027,000 convicted offender and arrestee samples in 2009 and an estimated 904,000 samples in 2014, but the difference in processed samples is not statistically significant.[44] However, there was a statistically significant decrease in requests for analysis of convicted offender and arrestee samples (1,053,000 requests were received in 2009 compared to 908,000 in 2014).[45]

FEDERAL LAW

While state law dictates which profiles will be included in each state's DNA database, federal law provides for the collection of DNA samples from certain federal offenders for analysis and inclusion in the NDIS. Federal law also dictates which profiles included in SDIS can be uploaded

---

[40] The difference between two estimated figures is considered to be statistically significant if the 95% confidence interval for the difference between the two figures does not contain zero. The confidence interval for the difference between two figures is equal to the difference plus and minus the pooled standard error (SE) multiplied by 1.96 (i.e., difference ± 1.96 * SE). BJS provides standard errors for the estimated figures in their report in the report's appendix.

[41] Ibid., Table 4.

[42] Ibid.

[43] Ibid., Table 6.

[44] Ibid., Table 4.

[45] Ibid.

into the NDIS. Federal law also states that agencies participating in the NDIS must meet certain specified standards. In addition, federal law provides for postconviction DNA testing for federal offenders. The following section summarizes current federal law as it pertains to DNA used in a criminal justice capacity.

Quality Assurance and Proficiency Testing Standards

Under current law,[46] the FBI is required to issue (and revise from time to time) Quality Assurance Standards (QAS), including standards for testing the proficiency of forensic laboratories and forensic analysts in conducting DNA analyses.[47] By law, the QAS must specify the criteria for quality assurance and proficiency tests to be applied to the various types of DNA analyses conducted by forensic laboratories.[48] The Rapid DNA Act of 2017 (Rapid DNA Act, P.L. 115-50) also requires the FBI to issue standards and procedures for the use of rapid DNA instruments and the resulting analyses.[49] The QAS must include a system for grading proficiency testing performance to determine whether a laboratory is performing acceptably.[50] Under current law, FBI personnel who perform DNA analyses must undergo semiannual external proficiency testing by a DNA proficiency testing program that meets the standards set in the QAS.[51]

According to the FBI, the QAS describe the minimum standards for a laboratory's quality assurance program if it is performing forensic DNA analysis and/or databasing.[52] The minimum standards cover the following areas: organization, personnel, facilities, evidence or sample control, validation, analytical procedures, equipment calibration and maintenance,

[46] 34 U.S.C. §12591(a)(2).
[47] The most recent QAS took effect on September 1, 2011.
[48] 34 U.S.C. §12591(a)(3).
[49] Rapid DNA, or rapid DNA analysis, is a term used to describe the fully automated process of developing a DNA profile from a cheek swab without human intervention.
[50] Ibid.
[51] 34 U.S.C. §12593(a)(1)(A).
[52] CODIS FAQs.

reports, review, proficiency testing, corrective action, audits, safety, and outsourcing.[53]

Index to Facilitate Law Enforcement Exchange of DNA Identification Information

The Violent Crime Control and Law Enforcement Act of 1994 (P.L. 103-322) authorized the FBI to establish an index of DNA profiles (i.e., NDIS). Under current law,[54] the NDIS can contain the DNA profiles of samples

- taken from individuals convicted of or charged with a crime, or collected under applicable legal authorities (e.g., people arrested for crimes), except for DNA samples that are voluntarily submitted solely for elimination purposes;
- recovered from crime scenes;
- recovered from unidentified human remains; and
- voluntarily contributed from relatives of missing persons.[55]

The NDIS can only include DNA profiles

- based on analyses performed by or on behalf of a criminal justice agency or the Department of Defense (DOD) in accordance with

---

[53] Ibid.

[54] 34 U.S.C. §12592(a).

[55] Under the Violent Crime Control and Law Enforcement Act of 1994 (P.L. 103-322), the NDIS was only to include analyses of DNA samples collected from (1) individuals convicted of crimes, (2) crime scenes, and (3) unidentified human remains. The Justice for All Act of 2004 (P.L. 108-405) amended the authorizing legislation for the NDIS to allow analyses of DNA samples collected from persons who have been charged in an indictment or information with a crime and other persons whose DNA samples are collected under applicable legal authorities to be included in the NDIS, provided that profiles from arrestees who have not been charged with a crime and samples that are voluntarily submitted solely for elimination purposes are not included in the NDIS. The Violence Against Women and Department of Justice Reauthorization Act of 2005 (P.L. 109-162) amended the authorizing legislation for the NDIS to allow analyses of samples collected from arrestees to be included in the NDIS.

available standards that satisfy or exceed the FBI's published QAS;

- that are prepared by laboratories that (1) have been accredited by a nonprofit professional organization of persons actively involved in forensic science and nationally recognized within the forensic science community, and (2) undergo external audits, not less than once every other year, that demonstrate compliance with the FBI's QAS;[56]
- that are prepared by criminal justice agencies using rapid DNA instruments approved by the FBI in compliance with the standards and procedures the FBI is required to publish per the Rapid DNA Act; and
- that are maintained by federal, state, and local criminal justice agencies or the DOD pursuant to rules that allow the disclosure of profiles only to other criminal justice agencies for identification purposes, judicial proceedings, criminal defense purposes, and, if personally identifiable information is removed, for research and quality control purposes.[57]

Under current law, the FBI is required to expunge the DNA profile of an individual who had a DNA profile entered into the NDIS on the basis of being convicted for a qualifying federal offense (see below) if the individual provides a certified copy of a final court order showing that the conviction was overturned.[58] Also, the FBI is required to expunge the DNA profile of an individual who had a DNA profile entered into the NDIS on the basis of being arrested under the authority of the United States if the individual provides a certified copy of a final court order that establishes that the charge was dismissed or resulted in an acquittal, or that no charge

---

[56] According to the FBI, American Association for Laboratory Accreditation (A2LA), and ANSI-ASQ National Accreditation Board (ANAB: The American Society of Crime Laboratory Directors/Laboratory Accreditation Board (ASCLD/LAB) and Forensic Quality Services (FQS), approved separately as accrediting agencies are now part of ANAB) are recognized as accrediting agencies. See CODIS FAQs.

[57] 34 U.S.C. §12592(b).

[58] 34 U.S.C. §12592(d)(1)(A)(i).

was filed within the applicable time period.[59] As a condition of having access to the NDIS, states must also have in place a procedure whereby the state will expunge a profile from the state's database based on the same conditions applicable to a profile being expunged from the NDIS.[60] Also, under current law the Department of Defense is required to expunge the DNA profile of an individual who had a DNA profile entered into the NDIS on the basis of being convicted of a qualifying military offense (see below) if the individual provides a certified copy of a final court order showing that the conviction was overturned.[61]

COLLECTION OF DNA SAMPLES FROM CERTAIN FEDERAL, DISTRICT OF COLUMBIA, AND MILITARY OFFENDERS

Under current law,[62] the Attorney General is permitted to collect DNA samples from "individuals who are arrested, facing charges, or convicted of a crime or from non-United States citizens who are detained under the authority of the United States."[63] In addition, the Bureau of Prisons (BOP)

---

[59] 34 U.S.C. §12592(d)(1)(A)(ii).

[60] 34 U.S.C. §12592(d)(2)(A)(i).

[61] 10 U.S.C. §1565(e).

[62] 34 U.S.C. §40702(a)(1)(A).

[63] The DNA Analysis Backlog Elimination Act of 2000 (P.L. 106-546) required BOP and U.S. probation offices to collect DNA samples from anyone in their custody who was convicted of qualifying federal offenses. The act defined a "qualifying federal offense" as murder, voluntary manslaughter, or other offenses relating to homicide; an offense relating to sexual abuse, sexual exploitation or other abuse of children, or transportation for illegal sexual activity; an offense relating to peonage or slavery; kidnapping; an offense relating to robbery or burglary; any offense committed in Indian country relating to murder, manslaughter, kidnapping, maiming, a felony sexual abuse offense, incest, arson, robbery, or burglary; or any attempt or conspiracy to commit any of these crimes. The Uniting and Strengthening America by Providing Appropriate Tools Required to Intercept and Obstruct Terrorism (USA PATRIOT) Act of 2001 (P.L. 107-56) expanded the definition of "qualifying federal offense" to include crimes of terrorism, crimes of violence, or any attempt or conspiracy to commit either crime. The Justice for All Act of 2004 (P.L. 108-405) amended the definition of "qualifying federal offense" to include any felony, sexual abuse offense, crime of violence, or attempt or conspiracy to commit any of these crimes. The Violence Against Women and Department of Justice Reauthorization Act of 2005 (P.L. 109-162) authorized DOJ to collect DNA samples from arrestees and non-citizens who are detained under the authority of the United States. The Adam Walsh Child Protection and

is required to collect a DNA sample from each federal prisoner who is, or has been, convicted of a felony, a sexual abuse crime under chapter 109A of title 18 of the U.S. Code, a crime of violence,[64] or any attempt or conspiracy to commit any of these crimes.[65] Federal probation offices responsible for supervising individuals on probation, parole, or supervised release are required to collect DNA samples from individuals who are, or have been, convicted of any of the crimes outlined above.[66] Collected samples are required to be submitted to the FBI for analysis and their resulting DNA profiles are included in the NDIS.[67] The Rapid DNA Act allows the FBI to waive the requirement that DNA samples be submitted to the FBI for analysis if the analysis is conducted with a rapid DNA instrument and the results are included in the NDIS.

Current law contains similar provisions regarding the collection of DNA samples from District of Columbia offenders. BOP is required to collect a DNA sample from each prisoner who is, or has been, convicted of a qualifying District of Columbia offense.[68] In addition, the Court Services and Offender Supervision Agency for the District of Columbia is required to collect DNA samples from individuals on probation, parole, or supervised release, who are, or have been, convicted of any qualifying District of Columbia offense.[69] The government of the District of Columbia may determine which offenses under the District of Columbia Code are considered qualifying offenses for the purposes of supplying a DNA sample.[70] Collected samples must be submitted to the FBI for analysis and their resulting DNA profiles are included in the NDIS.[71] The Rapid DNA

Safety Act of 2006 (P.L. 109-248) authorized DOJ to also collect DNA samples from individuals facing charges in addition to those who have been arrested or convicted.

[64] As defined at 18 U.S.C. §16.

[65] 34 U.S.C. §40702(a)(1)(B).

[66] 34 U.S.C. §40702(a)(2).

[67] 34 U.S.C. §40702(b).

[68] 34 U.S.C. §40703(a)(1).

[69] 34 U.S.C. §40703(a)(2).

[70] 34 U.S.C. §40703(d).

[71] 34 U.S.C. §40703(b). The following are considered qualifying offenses under the D.C. Code: (1) any felony; (2) any offense for which the penalty is greater than one year imprisonment; (3) lewd, indecent, or obscene acts knowingly committed in the presence of a child under 16 years of age (D.C. Code §22-1312(b)); (4) certain obscene activities involving minors (D.C. Code §22-2201); (5) sexual performances using a minor (D.C. Code §22-3102); (6)

Act also allows the FBI to waive the requirement that DNA samples be submitted to the FBI for analysis if the analysis is conducted with a rapid DNA instrument and the results are included in the NDIS for District of Columbia offenders.

Under current law,[72] the DOD is required to collect DNA samples from each member of the Armed Forces who is, or has been, convicted of an offense under the Uniform Code of Military Justice for which a sentence of confinement of more than one year can be imposed, or of any other offense under the Uniform Code of Military Justice that is comparable to the offenses for which a DNA sample can be collected from a federal offender (see above).[73] DOD is required to conduct an analysis of the collected sample and submit the results to the FBI for inclusion in the NDIS.[74]

Postconviction DNA Testing

The Justice for All Act of 2004 (P.L. 108-405, as amended) established procedures for postconviction DNA testing in federal courts. Under current law,[75] upon a written motion from an individual sentenced for a federal offense (hereinafter, "applicant"), the court must order DNA testing of evidence if all of the following apply:

- The applicant asserts, under penalty of perjury, that the applicant is actually innocent of the federal crime for which the applicant was sentenced, or another federal or state offense, if (1) "the evidence was entered during a federal death sentence hearing and exoneration for the offense would entitle the applicant to a reduced

misdemeanor sexual abuse (D.C. Code §22-3006); (7) misdemeanor sexual abuse of child or a minor (D.C. Code §22-3010.01); or (8) any attempt or conspiracy to commit any of these crimes. D.C. Code §22-4151.

[72] 10 U.S.C. §1565(a)(1).

[73] The requirement to collect DNA samples for people convicted of certain offenses under the Uniform Code of Military Justice is separate from the DNA samples the Department of Defense collects to aid in the identification of human remains.

[74] 10 U.S.C. §1565(b).

[75] 18 U.S.C. §3600(a).

sentence or a new sentencing hearing"; or (2) "in the case of a [s]tate offense, the applicant demonstrates that there is no adequate remedy under [s]tate law to permit DNA testing of the … evidence … and, to the extent available, the applicant has exhausted all remedies available under [s]tate law for requesting DNA testing of … evidence."

- The specified evidence to be tested was secured in relation to the investigation or prosecution of the federal or state crime for which the applicant claims to be innocent.
- The evidence to be tested (1) "was not previously subjected to DNA testing and the applicant did not knowingly fail to request DNA testing of that evidence in a prior motion for postconviction DNA testing"; or (2) "was previously subjected to DNA testing and the applicant requests DNA testing using a new method or technology that is substantially more probative that prior testing."
- The evidence to be tested "is in the possession of the [g]overnment and has been subject to a chain of custody and retained under conditions sufficient to ensure that such evidence has not been substituted, contaminated, tampered with, replaced, or altered in any respect" that would affect the DNA testing.
- The proposed DNA testing is "reasonable in scope, uses scientifically sound methods, and is consistent with accepted forensic practices."
- The applicant "identifies a theory of defense that is not inconsistent with an affirmative defense presented at trial and would establish the actual innocence of the applicant."
- If the applicant was "convicted following a trial, the identity of the perpetrator was at issue in the trial."
- The proposed DNA testing may produce new material evidence that would support the affirmative defense theory presented at trial and raise a reasonable probability that the applicant did not commit the crime.
- The applicant certifies that he or she will provide a DNA sample for comparison purposes.

- The motion is made in a timely fashion.[76]

If the court orders DNA testing, the testing is carried out by the FBI.[77] However, the court can order DNA testing to be conducted by another "qualified laboratory if the court makes all necessary orders to ensure the integrity of the … evidence and the reliability of the testing process and results."[78] The cost of any DNA testing is borne by the applicant, unless the applicant is indigent; in that case, the cost of the DNA testing is borne by the government.[79]

The results of any DNA test must be simultaneously provided to the court, applicant, and U.S. Attorney's office.[80] If the DNA test excludes the applicant as the source of the biological evidence, the DNA profile is required to be run through CODIS, assuming that the analysis was conducted in a manner consistent with the FBI's QAS, to see if the probative sample matches any profiles in the NDIS. The results of this search are to be simultaneously provided to the court, applicant, and U.S. Attorneys office. If the test results ordered by the court are "inconclusive or show that the applicant was the source of the tested evidence, the applicant's DNA profile may be retained in the NDIS."[81] Moreover, if the test results show that the applicant was not the source of the tested

[76] There is a rebuttable presumption of timeliness if the motion is made within 60 months of the enactment of the Justice for All Act of 2004 (October 30, 2004) or within 36 months of conviction, whichever comes later. The presumption of timeliness may be rebutted upon a showing that the applicant's motion for DNA testing is based solely upon information used in a previously denied motion or of clear and convincing evidence that the applicant's filing is done solely to cause delay or harass. For any motion that is not made within 60 months of the enactment of the Justice for All Act of 2004 or within 36 months of conviction, there is a rebuttable presumption against timeliness. The presumption against timeliness can be rebutted upon the court's finding (1) that the applicant was or is incompetent and such incompetence substantially contributed to the delay in the applicant's motion for a DNA test; (2) the evidence to be tested is newly discovered DNA evidence; (3) that the applicant's motion is not based solely upon the applicant's own assertion of innocence and, after considering all relevant facts and circumstances surrounding the motion, a denial would result in a manifest injustice; or (4) upon good cause shown. 18 U.S.C. §3600(a)(10)(B).

[77] 18 U.S.C. §3600(c)(1).

[78] 18 U.S.C. §3600(c)(2).

[79] 18 U.S.C. §3600(c)(3).

[80] 18 U.S.C. §3600(e)(1).

[81] 18 U.S.C. §3600(e)(3)(A).

evidence, and a comparison of the applicant's DNA profile with other forensic profiles in the NDIS results in a match, DOJ is to contact the appropriate agency and preserve the applicant's DNA sample.[82] However, if the test results exclude the applicant as the source of the tested evidence, and a comparison between the applicant's DNA profile and forensic profiles in the NDIS does not result in a match, DOJ must destroy the applicant's DNA sample and ensure that the applicant's DNA profile is not stored in the NDIS if there is no other legal authority to retain the profile in the NDIS.[83]

If the results of the DNA test are inconclusive, the court can order further testing, if appropriate, or it can deny the applicant relief.[84] If the results of the DNA test demonstrate that the applicant was the source of the evidence tested, the applicant is denied relief, and on a motion of the government, the court can determine whether the applicant's claim of actual innocence was false. If the court finds the claim was false, it can

- hold the applicant in contempt of court;
- assess against the applicant any cost of DNA testing;
- forward the findings to BOP, who may wholly, or in part, deny the applicant's good conduct time;[85]
- if the applicant is eligible for parole, forward the finding to the U.S. Parole Commission so the commission can deny parole on the basis of the finding; or
- if the test results relate to a state offense, forward the findings to the appropriate state official.[86]

Under current law, if the applicant is convicted of making false assertions relating to postconviction DNA testing, the applicant is to be

[82] 18 U.S.C. §3600(e)(3)(B).

[83] 18 U.S.C. §3600(e)(3)(C).

[84] 18 U.S.C. §3600(f)(1).

[85] Each prisoner serving a term of imprisonment of more than one year, but not prisoners serving a life sentence, can receive a good time credit of up to 54 days per year to count toward serving the sentence. The amount of the credit is subject to the determination of BOP. 18 U.S.C. §3624(b).

[86] 18 U.S.C. §3600(f)(2).

sentenced to no less than three years' imprisonment, to run consecutively with any other term of imprisonment the applicant is serving.[87]

If the results of the DNA testing demonstrate that the applicant was not the source of the tested evidence presented as a part of the case against the applicant, the applicant can file a motion for a new trial or resentencing, as appropriate, notwithstanding any law that would bar the motion as untimely.[88] Under current law, a court shall grant a motion for a new trial or resentencing, as appropriate, if DNA test results, when considered with all other evidence in the case (regardless of whether such evidence was introduced at trial), establish by compelling evidence that a new trial would result in an acquittal of

- in the case of a motion for a new trial, the federal offense for which the applicant is sentenced to imprisonment or death; and
- in the case of a motion for resentencing, another federal or state offense, if evidence of such offense was admitted during a federal sentencing hearing and exoneration for the offense would entitle the applicant to a reduced sentence or a new sentencing hearing.[89]

Preservation of Biological Evidence

The Justice for All Act of 2004 (P.L. 108-405, as amended), among other things, established standards for preserving biological evidence. Under current law,[90] the federal government is required to preserve biological evidence[91] that was secured in the investigation or prosecution of a federal offense, if a defendant was imprisoned for the offense, unless[92]

---

[87] 18 U.S.C. §3600(f)(3).
[88] 18 U.S.C. §3600(g)(1).
[89] 18 U.S.C. §3600(g)(2).
[90] 18 U.S.C. §3600A(a).
[91] "Biological evidence" is defined as a sexual assault forensic examination kit, or semen, blood, saliva, hair, skin tissue, or other identified biological material. 18 U.S.C. §3600A(b).
[92] 18 U.S.C. §3600A(c).

- "after a conviction becomes final and the defendant has exhausted all opportunities for direct review of the conviction, the defendant is notified that the evidence may be destroyed and the defendant does not file a motion [for post- conviction DNA testing] within 180 days of receipt of notice";
- "the evidence must be returned to its rightful owner, or it is of such size, bulk, or physical character as to render retention impracticable and the [g]overnment takes reasonable measures to remove and preserve portions of the evidence sufficient to permit future DNA testing"; or
- the evidence has been the subject of postconviction DNA testing (see above) and the results of the testing demonstrate that the defendant was the source of the evidence.

GRANTS FOR DNA-RELATED PROGRAMS

Several grant programs provide assistance to state and local governments for forensic sciences. A bulk of the programs focus on providing state and local governments with funding to reduce the backlog of forensic and convicted offender samples waiting to be processed and entered into the NDIS. However, some grant programs provide funding for other purposes, such as offsetting the cost of providing postconviction DNA testing. This section of the report provides a brief overview of grants for forensic sciences.

Debbie Smith DNA Backlog Grant Program

The Debbie Smith DNA Backlog Grant Program (hereinafter, "Debbie Smith grants") provides grants to state and local governments for five major purposes: (1) conducting analyses of DNA samples collected under applicable legal authority for inclusion in the NDIS, (2) conducting analyses of forensic DNA samples for inclusion in the NDIS, (3) increasing the capacity of state and local laboratories to carry out DNA analyses, (4) collecting DNA samples from people required to submit them and forensic samples from crimes, and (5) ensuring that analyses of forensic DNA samples are carried out in a timely manner. The Katie Sepich Enhanced DNA Collection Act of 2012 (P.L. 112-253) amended the Debbie Smith program to set aside up to $10 million of the amount appropriated for Debbie Smith grants for FY2013-FY2015 to assist states with the costs associated with collecting DNA samples from arrestees (assuming there is statutory authority in the state to collect DNA sample from people arrested for certain offenses). The Sexual Assault Forensic Evidence Reporting Act of 2013 (the SAFER Act of 2013, Title X of P.L. 113-4) added two new purposes for which Debbie Smith grants can be used: to conduct an audit of sexual assault evidence samples in the possession of a state or unit of local government that are awaiting testing and to ensure that the collection and processing of DNA evidence by law enforcement is carried out in a timely manner and in accordance with the protocols and practices the FBI is required to develop under the act.

The Attorney General is required to award these funds using a formula. The formula distributes funds among state and local governments to maximize the effective utilization of DNA technology to solve crimes and protect public safety. The formula must also allocate funding among state and local governments to reduce backlogs by considering the number of offender and forensic samples awaiting DNA analysis in the jurisdiction along with the population and number of violent crimes in the jurisdiction. Current law requires DOJ to award not less than 0.5% of the total amount appropriated each fiscal year to each state and the District of Columbia. The territories are to receive 0.125% of the total appropriation.

Agencies receiving a grant under the program are required to certify that DNA analyses are conducted in laboratories that satisfy the FBI's QAS and are operated either by a state or local government or by a private laboratory under contract with the state or local government. Grants for conducting analyses of DNA samples collected under applicable legal authority for inclusion in the NDIS, conducting analyses of forensic casework for inclusion in the NDIS, and ensuring that analyses of forensic

DNA samples are carried out in a timely manner can be made in the form of a contract or voucher for laboratory services that can be redeemed by nonprofit or for-profit laboratories that satisfy the QAS and have been approved by the Attorney General.

State and local governments receiving funding under the program are required to submit a report to DOJ with a summary of the activities carried out under the grant and an assessment of whether such activities are meeting the needs identified in the grant application, as well as other information the Attorney General may require. DOJ may award not more than 1% of grant funding each fiscal year to states, units of local government, and nonprofit professional organizations of persons actively involved in forensic science and nationally recognized within the forensic science community to help offset the cost of accrediting and auditing laboratories.

The SAFER Act of 2013 established a series of conditions for states or units of local government receiving a grant under the Debbie Smith program for the purposes of conducting an audit of sexual assault evidence. The act, among other things, requires states and local governments receiving grants for this purpose to (1) submit a plan for performing an audit of samples, (2) provide an estimate of the number of samples, (3) complete the audit within one year of receiving the grant, and (4) submit a report to DOJ every 60 days for at least one year after the audit is completed that provides data on the number of samples in the state's or unit of local government's possession along with data on new sexual assault evidence the state or local government receives and how those samples are being processed.

The SAFER Act of 2013 also requires the FBI, in consultation with federal, state, and local law enforcement agencies, to develop protocols and practices for the accurate, timely, and effective collection and processing of DNA evidence, including protocols and practices specific to sexual assault cases. The protocols and practices are required to address (1) what evidence should be collected by law enforcement and forwarded for testing and the order in which that evidence should be tested, (2) a reasonable period of time for evidence to be forwarded to a laboratory for testing, (3) a reasonable period of time in which each stage of laboratory testing should be conducted, (4) a system to encourage communication between actors in the criminal justice system (e.g., law enforcement, courts, and laboratory personnel and crime victims) about the status of evidence testing, and (5) standards for audits of sexual assault evidence in the possession of state and local governments.

Debbie Smith Grants were Originally Authorized under the Justice for All Act of 2004 (P.L. 108-405). This law amended the DNA Backlog Elimination Act of 2000,[93] authorizing appropriations of $151 million for each of FY2004-FY2009. The program was reauthorized under the Debbie Smith Reauthorization Act of 2008 (P.L. 110-360), which includes authorized appropriations of $151 million for FY2009-FY2014. The Debbie Smith Reauthorization Act of 2014 (P.L. 113-182) extended the $151 million per fiscal year authorization until FY2019.

Kirk Bloodsworth Post-Conviction DNA Testing Grant Program

The Kirk Bloodsworth DNA Post-Conviction DNA Testing Grant program was authorized by the Justice for All Act of 2004 (P.L. 108-405). The act authorized the Attorney General to make grants to states to help defray the costs of postconviction DNA testing programs. The act authorized appropriations of $5 million for FY2005-FY2009. The Justice for All Reauthorization Act of 2016 (P.L. 114-324) reauthorized appropriations for this program at $5 million per year for FY2017-FY2021.

Sexual Assault Forensic Exam Program Grants

The Sexual Assault Forensic Exam Program Grants were authorized under the Justice for All Act of 2004 (P.L. 108-405). The program

---

[93] The DNA Backlog Elimination Act of 2000 (P.L. 106-546) authorized grants to increase the capacity of state and local government laboratories to conduct DNA analysis of biological samples from crime scenes.

provides grants for training, technical assistance, education, equipment, and information relating to the identification, collection, preservation, analysis, and use of DNA samples and evidence by medical personnel and those treating victims of sexual assault. Under the program, entities eligible to receive grants include states, units of local government, and sexual assault examination programs. The act authorized appropriations of $30 million for each of FY2005-FY2009. P.L. 110-360 extended the same authorized amount through FY2014. The Debbie Smith Reauthorization Act of 2014 (P.L. 113-182) extended the $30 million per fiscal year authorization until FY2019.

DNA Research and Development Grants

The Justice for All Act of 2004 authorized grants for research and development for improving forensic DNA technology, including increasing the accuracy and efficiency of DNA analysis, decreasing the time and expense of conducting DNA analysis, and increasing its portability. In addition, the law authorized grants for demonstration projects to evaluate the use of DNA technology in conjunction with other forensic analyses. The act authorized funding of $15 million for each of FY2005-FY2009. The Justice for All Reauthorization Act of 2016 (P.L. 114-324) reauthorized appropriations for this program at $5 million per year for FY2017-FY2021.

DNA Training and Education for Law Enforcement, Correctional Personnel, and Court Officers

Under this program, the Attorney General is required to make grants to provide training, technical assistance, education, and information regarding the identification, collection, preservation, analysis, and use of DNA samples and evidence by law enforcement personnel, court officers, forensic science professionals, and corrections personnel. The program was originally authorized under the Justice for All Act of 2004 (P.L. 108-405), which authorized $12.5 million for each of FY2005-FY2009. P.L. 110-360 extended the same authorized amount through FY2014. The Debbie Smith Reauthorization Act of 2014 (P.L. 113-182) extended the $12.5 million per fiscal year authorization until FY2019.

In: DNA
Editor: Tomáš Koláček
ISBN: 978-1-53616-117-5

Chapter 2

RAPID DNA ACT OF 2017*

House of Representatives

Mr. GOODLATTE, from the Committee on the Judiciary, submitted the following report [To accompany H.R. 510] [Including cost estimate of the Congressional Budget Office].

The Committee on the Judiciary, to whom was referred the bill (H.R. 510) to establish a system for integration of Rapid DNA instruments for use by law enforcement to reduce violent crime and reduce the current DNA analysis backlog, having considered the same, reports favorably thereon without amendment and recommends that the bill do pass.

PURPOSE AND SUMMARY

H.R. 510 will establish a system for integration of Rapid DNA instruments for use by law enforcement to reduce violent crime and reduce the current DNA analysis backlog.

* This is an edited, reformatted and augmented version of House of Representatives, 115th Congress, 1st Session, Publication No. Report 115–117, dated May 11, 2017.

BACKGROUND AND NEED FOR THE LEGISLATION

The DNA Identification Act of 1994[1] established federal DNA labs and authorized the Federal Bureau of Investigation to begin compiling DNA information into a central database. This database is known as the National DNA Index System (NDIS) and the system for analyzing and communicating the data is the Combined DNA Index System (CODIS). The 1994 Act allowed DNA testing to be done by accredited state labs with results from state labs being uploaded to CODIS.

DNA technology has advanced a great deal in the years since the 1994 Act. Whereas it once took days or weeks, DNA testing can now be completed in a matter of hours. There is currently technology, known as Rapid DNA technology, that allows for DNA testing and identification on a small, copier-sized machine. A DNA sample—oftentimes a cheek swab—is taken, placed into a cartridge that slides into the Rapid DNA machine, and reports back the DNA profile in approximately ninety minutes. The FBI, working with the forensics community, is hopeful that this technology can be used in a booking station to help identify suspects in the same way a fingerprint is currently used. At present, Rapid DNA technology can only be used for identification purposes, not crime scene analysis.

Unfortunately, the 1994 Act creating CODIS does not allow for the use of this technology since only state labs are allowed access to CODIS. Currently, booking stations have to send their DNA samples off to state labs and wait weeks for the results. This has created a backlog that impacts all criminal investigations using forensics, not just forensics used for identification purposes. H.R. 510 would modify the current law regarding DNA testing and access to CODIS. The short turnaround time resulting from increased use of Rapid DNA technology would help to quickly eliminate potential suspects, capture those who have committed a previous crime and left DNA evidence, as well as free up current DNA profilers to

---

[1] Pub. L. No. 103–322 (1994).

do advanced forensic DNA analysis, such as crime scene analysis and rape-kits.

HEARINGS

The Committee on the Judiciary held no hearings on H.R. 510. However, the Judiciary Committee's Subcommittee on Crime, Terrorism, Homeland Security and Investigations held a hearing on a virtually identical bill, H.R. 320, on June 18, 2015. Testimony was received from: Ms. Amy Hess, Executive Assistant Director of Science and Technology, Federal Bureau of Investigation; Ms. Jody Wolf, Assistant Crime Laboratory Administrator, Phoenix Police Department Crime Laboratory, President, American Society of Criminal Laboratory Directors; and Ms. Natasha Alexenko, Founder, Natasha's Justice Project.

COMMITTEE CONSIDERATION

On April 27, 2017, the Committee met in open session and ordered the bill H.R. 510 favorably reported, by voice vote, a quorum being present.

COMMITTEE VOTES

In compliance with clause 3(b) of rule XIII of the Rules of the House of Representatives, the Committee advises that there were no recorded votes during the Committee's consideration of H.R. 510.

COMMITTEE OVERSIGHT FINDINGS

In compliance with clause 3(c)(1) of rule XIII of the Rules of the House of Representatives, the Committee advises that the findings and recommendations of the Committee, based on oversight activities under clause 2(b)(1) of rule X of the Rules of the House of Representatives, are incorporated in the descriptive portions of this chapter.

New Budget Authority and Tax Expenditures

Clause 3(c)(2) of rule XIII of the Rules of the House of Representatives is inapplicable because this legislation does not provide new budgetary authority or increased tax expenditures.

Congressional Budget Office Cost Estimate

In compliance with clause 3(c)(3) of rule XIII of the Rules of the House of Representatives, the Committee sets forth, with respect to the bill, H.R. 510, the following estimate and comparison prepared by the Director of the Congressional Budget Office under section 402 of the Congressional Budget Act of 1974:

May 9, 2017.
Hon. Bob Goodlatte,
Chairman.

Dear Mr. Chairman: The Congressional Budget Office has prepared the enclosed cost estimate for H.R. 510, the Rapid DNA Act of 2017.

If you wish further details on this estimate, we will be pleased to provide them. The CBO staff contact is Mark Grabowicz.

Sincerely,

Keith Hall.

Enclosure.
cc: Honorable John Conyers Jr. Ranking Member

H.R. 510—Rapid DNA Act of 2017

As ordered reported by the House Committee on the Judiciary on April 27, 2017.

H.R. 510 would direct the Federal Bureau of Investigation (FBI) to issue standards and procedures for the use of certain automated processes to analyze DNA samples. The agency is currently undertaking activities to prepare for increased use of such enhanced analyses. Based on information from the FBI, CBO estimates that implementing the bill would cost less than $500,000 annually; such spending would be subject to the availability of appropriated funds.

Enacting the legislation would not affect direct spending or revenues; therefore, pay-as-you-go procedures do not apply. CBO estimates that enacting H.R. 510 would not increase net direct spending or on-budget deficits in any of the four consecutive 10-year periods beginning in 2028.

H.R. 510 contains no intergovernmental or private-sector mandates as defined in the Unfunded Mandates Reform Act and would not affect the budgets of state, local, or tribal governments.

The CBO staff contact for this estimate is Mark Grabowicz. The estimate was approved by H. Samuel Papenfuss, Deputy Assistant Director for Budget Analysis.

DUPLICATION OF FEDERAL PROGRAMS

No provision of H.R. 510 establishes or reauthorizes a program of the Federal government known to be duplicative of another Federal program, a program that was included in any report from the Government Accountability Office to Congress pursuant to section 21 of Public Law 111 139, or a program related to a program identified in the most recent Catalog of Federal Domestic Assistance.

Disclosure of Directed Rule Makings

The Committee estimates that H.R. 510 specifically directs to be completed no specific rule makings within the meaning of 5 U.S.C. 551.

Performance Goals and Objectives

The Committee states that pursuant to clause 3(c)(4) of rule XIII of the Rules of the House of Representatives, H.R. 510 is designed to establish a system for integration of Rapid DNA instruments for use by law enforcement to reduce violent crime and reduce the current DNA analysis backlog.

Advisory on Earmarks

In accordance with clause 9 of rule XXI of the Rules of the House of Representatives, H.R. 510 does not contain any congressional earmarks, limited tax benefits, or limited tariff benefits as defined in clause 9(e), 9(f), or 9(g) of rule XXI.

Section-by-Section Analysis

The following discussion describes the bill as reported by the Committee.

- *Section 1. Short Title.* This section cites the short title of the bill as the "Rapid DNA Act of 2016."
- *Section 2. Rapid DNA Instruments.* This section requires the Director of the FBI to issue standards and procedures for the use of Rapid DNA instruments and resulting analyses. It also amends the DNA Identification Act of 1994 to include laboratories that are

independently accredited and undergo external audits not less than every two years.

- *Section 3. Conforming Amendments Relating to Collection of DNA Identification Information.* This section amends the DNA Analysis Backlog Elimination Act of 2000 by allowing the Director of the FBI to waive certain requirements under the Act if the DNA samples are analyzed by means of Rapid DNA instruments and the results are included in CODIS.

CHANGES IN EXISTING LAW MADE BY THE BILL, AS REPORTED

In compliance with clause 3(e) of rule XIII of the Rules of the House of Representatives, changes in existing law made by the bill, as reported, are shown as follows (existing law proposed to be omitted is enclosed in black brackets, new matter is printed in italics, and existing law in which no change is proposed is shown in roman):

DNA Identification Act of 1994

Title XXI—State and Local Law Enforcement

Subtitle C—DNA Identification

Sec. 210303. Quality Assurance and Proficiency Testing Standards

(a) PUBLICATION OF QUALITY ASSURANCE AND PROFICIENCY TESTING STANDARDS.—(1)(A) Not later than 180 days after the date of enactment of this Act, the Director of the Federal Bureau of Investigation shall appoint an advisory board on DNA quality assurance methods from among nominations proposed by the head of the National Academy of Sciences and professional societies of crime laboratory officials.

(B) The advisory board shall include as members scientists from State, local, and private forensic laboratories, molecular geneticists and population geneticists not affiliated with a forensic laboratory, and a representative from the National Institute of Standards and Technology.

(C) The advisory board shall develop, and if appropriate, periodically revise, recommended standards for quality assurance, including standards for testing the proficiency of forensic laboratories, and forensic analysts, in conducting analyses of DNA.

(2) The Director of the Federal Bureau of Investigation, after taking into consideration such recommended standards, shall issue (and revise from time to time) standards for quality assurance, including standards for testing the proficiency of forensic laboratories, and forensic analysts, in conducting analyses of DNA.

(3) The standards described in paragraphs (1) and (2) shall specify criteria for quality assurance and proficiency tests to be applied to the various types of DNA analyses used by forensic laboratories. The standards shall also include a system for grading proficiency testing performance to determine whether a laboratory is performing acceptably.

(4) Until such time as the advisory board has made recommendations to the Director of the Federal Bureau of Investigation and the Director has acted upon those recommendations, the quality assurance guidelines adopted by the technical working group on DNA analysis methods shall be deemed the Director's standards for purposes of this section.

(5)(A) In addition to issuing standards as provided in paragraphs (1) through (4), the Director of the Federal Bureau of Investigation shall issue standards and procedures for the use of Rapid DNA instruments and resulting DNA analyses.

(B) In this Act, the term ''Rapid DNA instruments" means instrumentation that carries out a fully automated process to derive a DNA analysis from a DNA sample.

(b) ADMINISTRATION OF THE ADVISORY BOARD.—(1) For administrative purposes, the advisory board appointed under subsection (a) shall be considered an advisory board to the Director of the Federal Bureau of Investigation.

(2) Section 14 of the Federal Advisory Committee Act (5 U.S.C. App.) shall not apply with respect to the advisory board appointed under subsection (a).

(3) The DNA advisory board established under this section shall be separate and distinct from any other advisory board administered by the FBI, and is to be administered separately.

(4) The board shall cease to exist on the date 5 years after the initial appointments are made to the board, unless the existence of the board is extended by the Director of the Federal Bureau of Investigation.

(c) PROFICIENCY TESTING PROGRAM.—(1) Not later than 1 year after the effective date of this Act, the Director of the National Institute of Justice shall certify to the Committees on the Judiciary of the House and Senate that—

(A) the Institute has entered into a contract with, or made a grant to, an appropriate entity for establishing, or has taken other appropriate action to ensure that there is established, not later than 2 years after the date of enactment of this Act, a blind external proficiency testing program for DNA analyses, which shall be available to public and private laboratories performing forensic DNA analyses;

(B) a blind external proficiency testing program for DNA analyses is already readily available to public and private laboratories performing forensic DNA analyses; or

(C) it is not feasible to have blind external testing for DNA forensic analyses.

(2) As used in this subsection, the term "blind external proficiency test" means a test that is presented to a forensic laboratory

through a second agency and appears to the analysts to involve routine evidence.

(3) Notwithstanding any other provision of law, the Attorney General shall make available to the Director of the National Institute of Justice during the first fiscal year in which funds are distributed under this subtitle up to $250,000 from the funds available under part X of Title I of the Omnibus Crime Control and Safe Streets Act of 1968 to carry out this subsection.

Sec. 210304. Index to Facilitate Law Enforcement Exchange of DNA Identification Information

(a) ESTABLISHMENT OF INDEX.—The Director of the Federal Bureau of Investigation may establish an index of—

(1) DNA identification records of—

(A) persons convicted of crimes;

(B) persons who have been charged in an indictment or information with a crime; and

(C) other persons whose DNA samples are collected under applicable legal authorities, provided that DNA samples that are voluntarily submitted solely for elimination purposes shall not be included in the National DNA Index System;

(2) analyses of DNA samples recovered from crime scenes;

(3) analyses of DNA samples recovered from unidentified human remains; and

(4) analyses of DNA samples voluntarily contributed from relatives of missing persons.

(b) INFORMATION.—The index described in subsection (a) shall include only information on DNA identification records and DNA analyses that are—

(1) based on analyses performed by or on behalf of a criminal justice agency (or the Secretary of Defense in accordance with section 1565 of title 10, United States Code) in accordance with publicly

available standards that satisfy or exceed the guidelines for a quality assurance program for DNA analysis, issued by the Director of the Federal Bureau of Investigation under section 210303;

[(2) prepared by laboratories that—not later than 2 years after the date of enactment of the DNA Sexual Assault Justice Act of 2004, have been accredited by a nonprofit professional association of persons actively involved in forensic science that is nationally recognized within the forensic science community; and undergo external audits, not less than once every 2 years, that demonstrate compliance with standards established by the Director of the Federal Bureau of Investigation; and]

(2) prepared by—

(A) laboratories that—

(i) *have been accredited by a nonprofit professional association of persons actively involved in forensic science that is nationally recognized within the forensic science community; and*

(ii) *undergo external audits, not less than once every 2 years, that demonstrate compliance with standards established by the Director of the Federal Bureau of Investigation; or*

(B) criminal justice agencies using Rapid DNA instruments approved by the Director of the Federal Bureau of Investigation in compliance with the standards and procedures issued by the Director under section 210303(a)(5); and

(3) maintained by Federal, State, and local criminal justice agencies (or the Secretary of Defense in accordance with section 1565 of title 10, United States Code) pursuant to rules that allow disclosure of stored DNA samples and DNA analyses only—

(A) to criminal justice agencies for law enforcement identification purposes;

(B) in judicial proceedings, if otherwise admissible pursuant to applicable statutes or rules;

(C) for criminal defense purposes, to a defendant, who shall have access to samples and analyses performed in connection with the case in which such defendant is charged; or

(D) if personally identifiable information is removed, for a population statistics database, for identification research and protocol development purposes, or for quality control purposes.

(c) FAILURE TO COMPLY.—Access to the index established by this section is subject to cancellation if the quality control and privacy requirements described in subsection (b) are not met.

(d) EXPUNGEMENT OF RECORDS.—

(1) BY DIRECTOR.—

(A) The Director of the Federal Bureau of Investigation shall promptly expunge from the index described in subsection (a) the DNA analysis of a person included in the index—

(i) on the basis of conviction for a qualifying Federal offense or a qualifying District of Columbia offense (as determined under sections 3 and 4 of the DNA Analysis Backlog Elimination Act of 2000 (42 U.S.C. 14135a, 14135b), respectively), if the Director receives, for each conviction of the person of a qualifying offense, a certified copy of a final court order establishing that such conviction has been overturned; or

(ii) on the basis of an arrest under the authority of the United States, if the Attorney General receives, for each charge against the person on the basis of which the analysis was or could have been included in the index, a certified copy of a final court order establishing that such charge has been dismissed or has resulted in an acquittal or that no charge was filed within the applicable time period.

(B) For purposes of subparagraph (A), the term "qualifying offense" means any of the following offenses:

(i) A qualifying Federal offense, as determined under section 3 of the DNA Analysis Backlog Elimination Act of 2000.

(ii) A qualifying District of Columbia offense, as determined under section 4 of the DNA Analysis Backlog Elimination Act of 2000.

(iii) A qualifying military offense, as determined under section 1565 of title 10, United States Code.

(C) For purposes of subparagraph (A), a court order is not "final" if time remains for an appeal or application for discretionary review with respect to the order.

(2) BY STATES.—(A) As a condition of access to the index described in subsection (a), a State shall promptly expunge from that index the DNA analysis of a person included in the index by that State if—

(i) the responsible agency or official of that State receives, for each conviction of the person of an offense on the basis of which that analysis was or could have been included in the index, a certified copy of a final court order establishing that such conviction has been overturned; or

(ii) the person has not been convicted of an offense on the basis of which that analysis was or could have been included in the index, and the responsible agency or official of that State receives, for each charge against the person on the basis of which the analysis was or could have been included in the index, a certified copy of a final court order establishing that such charge has been dismissed or has resulted in an acquittal or that no charge was filed within the applicable time period.

(B) For purposes of subparagraph (A), a court order is not "final" if time remains for an appeal or application for discretionary review with respect to the order.

DNA Analysis Backlog Elimination Act of 2000

Sec. 3. Collection and Use of DNA Identification Information from Certain Federal Offenders

(a) COLLECTION OF DNA SAMPLES.—

(1) FROM INDIVIDUALS IN CUSTODY.—

(A) The Attorney General may, as prescribed by the Attorney General in regulation, collect DNA samples from individuals who are arrested, facingcharges, or convicted or from non-United States persons who are detained under the authority of the United States. The Attorney General may delegate this function within the Department of Justice as provided in section 510 of title 28, United States Code, and may also authorize and direct any other agency of the United States that arrests or detains individuals or supervises individuals facing charges to carry out any function and exercise any power of the Attorney General under this section.

(B) The Director of the Bureau of Prisons shall collect a DNA sample from each individual in the custody of the Bureau of Prisons who is, or has been, convicted of a qualifying Federal offense (as determined under subsection (d)) or a qualifying military offense, as determined under section 1565 of title 10, United States Code.

(2) FROM INDIVIDUALS ON RELEASE, PAROLE, OR PROBATION.—

The probation office responsible for the supervision under Federal law of an individual on probation, parole, or supervised release shall collect a DNA sample from each such individual who is, or has been, convicted of a qualifying Federal offense (as determined under subsection (d)) or a qualifying military offense, as determined under section 1565 of title 10, United States Code.

(3) INDIVIDUALS ALREADY IN CODIS.—For each individual described in paragraph (1) or (2), if the Combined DNA Index System (in this section referred to as "CODIS") of the Federal Bureau of Investigation contains a DNA analysis with respect to

that individual, or if a DNA sample has been collected from that individual under section 1565 of title 10, United States Code, the Attorney General, the Director of the Bureau of Prisons, or the probation office responsible (as applicable) may (but need not) collect a DNA sample from that individual.

(4) COLLECTION PROCEDURES.—(A) The Attorney General, the Director of the Bureau of Prisons, or the probation office responsible (as applicable) may use or authorize the use of such means as are reasonably necessary to detain, restrain, and collect a DNA sample from an individual who refuses to cooperate in the collection of the sample.

(B) The Attorney General, the Director of the Bureau of Prisons, or the probation office, as appropriate, may enter into agreements with units of State or local government or with private entities to provide for the collection of the samples described in paragraph (1) or (2).

(5) CRIMINAL PENALTY.—An individual from whom the collection of a DNA sample is authorized under this subsection who fails to cooperate in the collection of that sample shall be—

(A) guilty of a class A misdemeanor; and

(B) punished in accordance with title 18, United States Code.

(b) ANALYSIS AND USE OF SAMPLES.—The Attorney General, the Director of the Bureau of Prisons, or the probation office responsible (as applicable) shall furnish each DNA sample collected under subsection (a) to the Director of the Federal Bureau of Investigation, who shall carry out a DNA analysis on each such DNA sample and include the results in CODIS. *The Director of the Federal Bureau of Investigation may waive the requirements under this subsection if DNA samples are analyzed by means of Rapid DNA instruments and the results are included in CODIS.*

(c) DEFINITIONS.—In this section:

(1) The term "DNA sample" means a tissue, fluid, or other bodily sample of an individual on which a DNA analysis can be carried out.

(2) The term "DNA analysis" means analysis of the deoxyribonucleic acid (DNA) identification information in a bodily sample.

(3) *The term "Rapid DNA instruments" means instrumentation that carries out a fully automated process to derive a DNA analysis from a DNA sample.*

(d) QUALIFYING FEDERAL OFFENSES.—The offenses that shall be treated for purposes of this section as qualifying Federal offenses are the following offenses, as determined by the Attorney General:

(1) Any felony.

(2) Any offense under chapter 109A of title 18, United States Code.

(3) Any crime of violence (as that term is defined in section 16 of title 18, United States Code).

(4) Any attempt or conspiracy to commit any of the offenses in paragraphs (1) through (3).

(e) REGULATIONS.—

(1) IN GENERAL.—Except as provided in paragraph (2), this section shall be carried out under regulations prescribed by the Attorney General.

(2) PROBATION OFFICERS.—The Director of the Administrative Office of the United States Courts shall make available model procedures for the activities of probation officers in carrying out this section.

(f) COMMENCEMENT OF COLLECTION.—Collection of DNA samples under subsection (a) shall, subject to the availability of appropriations, commence not later than the date that is 180 days after the date of the enactment of this Act.

Sec. 4. Collection and Use of DNA Identification Information from Certain District of Columbia Offenders.

(a) COLLECTION OF DNA SAMPLES.—

(1) FROM INDIVIDUALS IN CUSTODY.—The Director of the Bureau of Prisons shall collect a DNA sample from each individual in the custody of the Bureau of Prisons who is, or has

been, convicted of a qualifying District of Columbia offense (as determined under subsection (d)).

(2) FROM INDIVIDUALS ON RELEASE, PAROLE, OR PROBATION.— The Director of the Court Services and Offender Supervision Agency for the District of Columbia shall collect a DNA sample from each individual under the supervision of the Agency who is on supervised release, parole, or probation who is, or has been, convicted of a qualifying District of Columbia offense (as determined under subsection (d)).

(3) INDIVIDUALS ALREADY IN CODIS.—For each individual described in paragraph (1) or (2), if the Combined DNA Index System (in this section referred to as "CODIS") of the Federal Bureau of Investigation contains a DNA analysis with respect to that individual, the Director of the Bureau of Prisons or Agency (as applicable) may (but need not) collect a DNA sample from that individual.

(4) COLLECTION PROCEDURES.—(A) The Director of the Bureau of Prisons or Agency (as applicable) may use or authorize the use of such means as are reasonably necessary to detain, restrain, and collect a DNA sample from an individual who refuses to cooperate in the collection of the sample.

(B) The Director of the Bureau of Prisons or Agency, as appropriate, may enter into agreements with units of State or local government or with private entities to provide for the collection of the samples described in paragraph (1) or (2).

(5) CRIMINAL PENALTY.—An individual from whom the collection of a DNA sample is authorized under this subsection who fails to cooperate in the collection of that sample shall be—

(A) guilty of a class A misdemeanor; and

(B) punished in accordance with title 18, United States Code.

(b) ANALYSIS AND USE OF SAMPLES.—The Director of the Bureau of Prisons or Agency (as applicable) shall furnish each DNA sample collected under subsection (a) to the Director of the Federal Bureau of Investigation, who shall carry out a DNA analysis on each such DNA sample and include the results in CODIS. *The Director of the Federal Bureau of Investigation may waive the requirements under this subsection if DNA samples are analyzed by means of Rapid DNA instruments and the results are included in CODIS.*

(c) DEFINITIONS.—In this section:

(1) The term "DNA sample" means a tissue, fluid, or other bodily sample of an individual on which a DNA analysis can be carried out.

(2) The term "DNA analysis" means analysis of the deoxyribonucleic acid (DNA) identification information in a bodily sample.

(3) *The term "Rapid DNA instruments" means instrumentation that carries out a fully automated process to derive a DNA analysis from a DNA sample.*

(d) QUALIFYING DISTRICT OF COLUMBIA OFFENSES.—The government of the District of Columbia may determine those offenses under the District of Columbia Code that shall be treated for purposes of this section as qualifying District of Columbia offenses.

(e) COMMENCEMENT OF COLLECTION.—Collection of DNA samples under subsection (a) shall, subject to the availability of appropriations, commence not later than the date that is 180 days after the date of the enactment of this Act.

(f)) Authorization of Appropriations.—There are authorized to be appropriated to the Court Services and Offender Supervision Agency for the District of Columbia to carry out this section such sums as may be necessary for each of fiscal years 2001 through 2005.

In: DNA
Editor: Tomáš Koláček
ISBN: 978-1-53616-117-5

Chapter 3

PROVIDING FOR CONSIDERATION OF THE BILL (S. 139) TO IMPLEMENT THE USE OF RAPID DNA INSTRUMENTS TO INFORM DECISIONS ABOUT PRETRIAL RELEASE OR DETENTION AND THEIR CONDITIONS, TO SOLVE AND PREVENT VIOLENT CRIMES AND OTHER CRIMES, TO EXONERATE THE INNOCENT, TO PREVENT DNA ANALYSIS BACKLOGS, AND FOR OTHER PURPOSES*

House of Representatives

Mr. COLLINS of Georgia, from the Committee on Rules, submitted the following report [To accompany H. Res. 682]

* This is an edited, reformatted and augmented version of 115th Congress, 2d Session, House of Representatives, Publication No. Report 115–504, dated January 9, 2018.

The Committee on Rules, having had under consideration House Resolution 682, by a record vote of 6 to 3, report the same to the House with the recommendation that the resolution be adopted.

Summary of Provisions of the Resolution

The resolution provides for consideration of S. 139, the Rapid DNA Act of 2017, under a structured rule. The resolution provides one hour of debate with 40 minutes equally divided and controlled by the chair and ranking minority member of the Permanent Select Committee on Intelligence, and 20 minutes equally divided and controlled by the chair and ranking minority member of the Committee on the Judiciary. The resolution waives all points of order against consideration of the bill. The resolution provides that an amendment in the nature of a substitute consisting of the text of Rules Committee Print 115–53 shall be considered as adopted and the bill, as amended, shall be considered as read. The resolution waives all points of against provisions in the bill, as amended. The resolution makes in order only the further amendment printed in this chapter, if offered by the Member designated in this chapter, which shall be considered as read, shall be debatable for the time specified in this chapter equally divided and controlled by the proponent and an opponent, shall not be subject to amendment, and shall not be subject to a demand for division of the question. The resolution waives all points of order against the amendment printed in this chapter. The resolution provides one motion to commit with or without instructions.

Explanation of Waivers

Although the resolution waives all points of order against consideration of the bill, the Committee is not aware of any points of order. The waiver is prophylactic in nature.

Although the resolution waives all points of order against provisions in the bill, as amended, the Committee is not aware of any points of order. The waiver is prophylactic in nature.

The waiver of all points of order against the amendment printed in this chapter includes a waiver of clause 12(a)(2) of rule XXI, which prohibits consideration of an amendment in the nature of a substitute unless there is a searchable electronic comparative print that shows how the amendment proposes to change current law.

COMMITTEE VOTES

The results of each record vote on an amendment or motion to report, together with the names of those voting for and against, are printed below:

Rules Committee Record Vote No. 173

Motion by Mr. Polis to make in order and provide the appropriate waivers to amendment #3, offered by Rep. Poe (TX) and Rep. Nadler (NY) and Rep. Lofgren (CA) and Rep. Gabbard (HI) and Rep. Jayapal (WA) and Rep. Gomez (CA) and Rep. Farenthold (TX) and Rep. Polis (CO), which mandates a warrant requirement for a search of the 702 database on a US person. It also ends "about" collection. Defeated: 3–6

Majority Members	**Vote**	**Minority Members**	**Vote**
Mr. Cole	Nay	Ms. Slaughter	Yea
Mr. Woodall	Nay	Mr. McGovern	Yea
Mr. Burgess		Mr. Hastings of Florida	
Mr. Collins	Nay	Mr. Polis	Yea
Mr. Byrne			
Mr. Newhouse	Nay		
Mr. Buck			
Ms. Cheney	Nay		
Mr. Sessions, Chairman	Nay		

Rules Committee Record Vote No. 174

Motion by Mr. Cole to report the rule. Adopted: 6–3

Majority Members	Vote	Minority Members	Vote
Mr. Cole	Yea	Ms. Slaughter	Nay
Mr. Woodall	Yea	Mr. McGovern	Nay
Mr. Burgess	Yea	Mr. Hastings of Florida	
Mr. Collins	Yea	Mr. Polis	Nay
Mr. Byrne			
Mr. Newhouse	Yea		
Mr. Buck			
Ms. Cheney			
Mr. Sessions, Chairman	Yea		

SUMMARY OF THE AMENDMENT IN THE NATURE OF A SUBSTITUTE MADE IN ORDER

1. Amash (MI): Replaces the text of S. 139 with the text of the USA RIGHTS Act, to protect Americans' Fourth Amendment rights. (20 minutes)

Text of Amendment in the Nature of a Substitute Made in Order

1. AN AMENDMENT IN THE NATURE OF A SUBSTITUTE TO BE OFFERED BY REPRESENTATIVE AMASH OF MICHIGAN OR HIS DESIGNEE, DEBATABLE FOR 20 MINUTES

Page 1, strike line 1 and all that follows and insert the following:

Section 1. Short Title; Table of Contents

(a) SHORT TITLE.—This Act may be cited as the "Uniting and Strengthening America by Reforming and Improving the

Government's High-Tech Surveillance Act" or the "USA RIGHTS Act."

(b) TABLE OF CONTENTS.—The table of contents for this Act is as follows:

- Sec. 12. Clarification of applicability of requirement to declassify significant decisions of Foreign Intelligence Surveillance Court and Foreign Intelligence Surveillance Court of Review.
- Sec. 13. Clarification regarding treatment of information acquired under Foreign Intelligence Surveillance Act of 1978.
- Sec. 14. Limitation on technical assistance from electronic communication service providers under the Foreign Intelligence Surveillance Act of 1978.
- Sec. 15. Modification of authorities for public reporting by persons subject to nondisclosure requirement accompanying order under Foreign Intelligence Surveillance Act of 1978.
- Sec. 16. Annual publication of statistics on number of persons targeted outside the United States under certain Foreign Intelligence Surveillance Act of 1978 authority.
- Sec. 17. Repeal of nonapplicability to Federal Bureau of Investigation of certain reporting requirements under Foreign Intelligence Surveillance Act of 1978.
- Sec. 18. Publication of estimates regarding communications collected under certain provision of Foreign Intelligence Surveillance Act of 1978.
- Sec. 19. Four-year extension of FISA Amendments Act of 2008.

Sec. 2. Clarification on Prohibition on Querying of Collections of Communications to Conduct Warrantless Queries for the Communications of United States Persons and Persons inside the United States

Section 702(b) of the Foreign Intelligence Surveillance Act of 1978 (50 U.S.C. 1881a(b)) is amended—

(1) by redesignating paragraphs (1) through (5) as subparagraphs (A) through (E), respectively, and indenting such subparagraphs, as so redesignated, an additional two ems from the left margin;

(2) by striking "An acquisition" and inserting the following:

"(1) IN GENERAL.—An acquisition"; and

(3) by adding at the end the following:

"(2) CLARIFICATION ON PROHIBITION ON QUERYING OF COLLECTIONS OF COMMUNICATIONS OF UNITED STATES PERSONS AND PERSONS INSIDE THE UNITED STATES.—

"(A) IN GENERAL.—Except as provided in subparagraphs (B) and (C), no officer or employee of the United States may conduct a query of information acquired under this section in an effort to find communications of or about a particular United States person or a person inside the United States.

"(B) CONCURRENT AUTHORIZATION AND EXCEPTION FOR EMERGENCY SITUATIONS.—Subparagraph (A) shall not apply to a query for communications related to a particular United States person or person inside the United States if—

(i) such United States person or person inside the United States is the subject of an order or emergency authorization authorizing electronic surveillance or physical search under section 105, 304, 703, 704, or 705 of this Act, or under title 18, United States Code, for the effective period of that order;

(ii) the entity carrying out the query has a reasonable belief that the life or safety of such United States person or person inside the United States is threatened and the information is sought for the purpose of assisting that person;

(iii) such United States person or person in the United States is a corporation; or

(iv) such United States person or person inside the United States has consented to the query.

"(C) QUERIES OF FEDERATED DATA SETS AND MIXED DATA.—If an officer or employee of the United States conducts a query of a data set, or of federated data sets, that includes any information acquired under this section, the system shall be configured not to return such information unless the officer or employee enters a code or other information indicating that—

(i) the person associated with the search term is not a United States person or person inside the United States; or

(ii) if the person associated with the search term is a United States person or person inside the United States, one or more of the conditions of subparagraph (B) are satisfied.

"(D) MATTERS RELATING TO EMERGENCY QUERIES.—

"(i) TREATMENT OF DENIALS.—In the event that a query for communications related to a particular United States person or a person inside the United States is conducted pursuant to an emergency authorization authorizing electronic surveillance or a physical search described in subsection (B)(i) and the application for such emergency authorization is denied, or in any other case in which the query has been conducted and no order is issued approving the query—

(I) no information obtained or evidence derived from such query may be received in evidence or otherwise disclosed in any trial, hearing, or other proceeding in or before any court, grand jury, department, office, agency, regulatory body, legislative committee, or other authority of the United States, a State, or political subdivision thereof; and

(II) no information concerning any United States person acquired from such query may subsequently be used or disclosed in any other manner by Federal officers or employees without the consent of such person, except with the approval of the Attorney General if the information indicates a threat of death or serious bodily harm to any person.

"(ii) ASSESSMENT OF COMPLIANCE.—The Attorney General shall assess compliance with the requirements under clause (i)."

Sec. 3. Prohibition on Reverse Targeting under Certain Authorities of the Foreign Intelligence Surveillance Act of 1978

Section 702 of the Foreign Intelligence Surveillance Act of 1978 (50 U.S.C. 1881a), as amended by section 2, is further amended—

(1) in subsection (b)(1)(B), as redesignated by section 2, by striking "the purpose of such acquisition is to target" and inserting "a significant purpose of such acquisition is to acquire the communications of";

(2) in subsection (d)(1)(A)—

(A) by striking "ensure that" and inserting the following: "ensure—"(i) that"; and

(B) by adding at the end the following: "(ii) that an application is filed under title I, if otherwise required, when a significant purpose of an acquisition authorized under subsection (a) is to acquire the communications of a particular, known person reasonably believed to be located in the United States; and";

(3) in subsection (g)(2)(A)(i)(I)—

(A) by striking "ensure that" and inserting the following: "ensure—"(aa) that"; and

(B) by adding at the end the following: "(bb) that an application is filed under title I, if otherwise required, when a significant purpose of an acquisition authorized under subsection (a) is to acquire the communications of a particular, known person reasonably believed to be located in the United States; and"; and

(4) in subsection (i)(2)(B)(i)—

(A) by striking "ensure that" and inserting the following: "ensure—"(I) that"; and

(B) by adding at the end the following: "(II) that an application is filed under title I, if otherwise required, when a significant purpose of an acquisition authorized under subsection (a) is to acquire the communications of a particular, known person reasonably believed to be located in the United States; and."

Sec. 4. Prohibition on Acquisition, Pursuant to Certain FISA Authorities to Target Certain Persons outside the United States, of Communications That Do Not Include Persons Targeted under Such Authorities

Section 702(b)(1) of the Foreign Intelligence Surveillance Act of 1978, as redesignated by section 2, is amended—

(1) in subparagraph (D), as redesignated by section 2, by striking "; and" and inserting a semicolon;
(2) by redesignating subparagraph (E) as subparagraph (G); and
(3) by inserting after subparagraph (D) the following:
(E) may not acquire a communication as to which no participant is a person who is targeted pursuant to the authorized acquisition."

Sec. 5. Prohibition on Acquisition of Entirely Domestic Communications under Authorities to Target Certain Persons outside the United States

Section 702(b)(1) of the Foreign Intelligence Surveillance Act of 1978, as redesignated by section 2 and amended by section 4, is further amended by inserting after subparagraph (E), as added by section 4, the following:

(F) may not acquire communications known to be entirely domestic; and."

Sec. 6. Limitation on Use of Information Obtained under Certain Authority of Foreign Intelligence Surveillance Act of 1947 Relating to United States Persons

Section 706(a) of the Foreign Intelligence Surveillance Act of 1978 (50 U.S.C. 1881e(a)) is amended—

(1) by striking "Information acquired" and inserting the following: "(1) IN GENERAL.—Information acquired"; and
(2) by adding at the end the following: "(2) LIMITATION ON USE IN CRIMINAL, CIVIL, AND ADMINISTRATIVE PROCEEDINGS

AND INVESTIGATIONS.—No communication to or from, or information about, a person acquired under section 702 who is either a United States person or is located in the United States may be introduced as evidence against the person in any criminal, civil, or administrative proceeding or used as part of any criminal, civil, or administrative investigation, except—

(A) with the prior approval of the Attorney General; and

(B) in a proceeding or investigation in which the information is directly related to and necessary to address a specific threat of—

(i) terrorism (as defined in clauses (i) through (iii) of section 2332(g)(5)(B) of title 18, United States Code);

(ii) espionage (as used in chapter 37 of title 18, United States Code);

(iii) proliferation or use of a weapon of mass destruction (as defined in section 2332a(c) of title 18, United States Code);

(iv) a cybersecurity threat from a foreign country;

(v) incapacitation or destruction of critical infrastructure (as defined in section 1016(e) of the USA PATRIOT Act (42 U.S.C. 5195c(e))); or

(vi) a threat to the armed forces of the United States or an ally of the United States or to other personnel of the United States Government or a government of an ally of the United States."

Sec. 7. Reforms of the Privacy and Civil Liberties Oversight Board.

(a) INCLUSION OF FOREIGN INTELLIGENCE ACTIVITIES IN OVERSIGHT AUTHORITY OF THE PRIVACY AND CIVIL LIBERTIES OVERSIGHT BOARD.—Section 1061 of the Intelligence Reform and Terrorism Prevention Act of 2004 (42 U.S.C. 2000ee) is amended—

(1) in subsection (c), by inserting "and to conduct foreign intelligence activities" after "terrorism" each place such term appears; and

(2) in subsection (d), by inserting "and to conduct foreign intelligence activities" after "terrorism" each place such term appears.

(b) SUBMISSION OF WHISTLEBLOWER COMPLAINTS TO THE PRIVACY AND CIVIL LIBERTIES OVERSIGHT BOARD.—

(1) IN GENERAL.—Section 1061 of the Intelligence Reform and Terrorism Prevention Act of 2004 (42 U.S.C. 2000ee), as amended by subsection (a), is further amended—

(A) in subsection (d), by adding at the end the following: "(5) WHISTLEBLOWER COMPLAINTS.—

"(A) SUBMISSION TO BOARD.—An employee of, or contractor or detailee to, an element of the intelligence community may submit to the Board a complaint or information that such employee, contractor, or detailee believes relates to a privacy or civil liberties concern. The confidentiality provisions under section 2409(b)(3) of title 10, United States Code, shall apply to a submission under this subparagraph. Any disclosure under this subparagraph shall be protected against discrimination under the procedures, burdens of proof, and remedies set forth in section 2409 of such title.

(B) AUTHORITY OF BOARD.—The Board may take such action as the Board considers appropriate with respect to investigating a complaint or information submitted under subparagraph (A) or transmitting such complaint or information to any other Executive agency or the congressional intelligence committees.

(C) RELATIONSHIP TO EXISTING LAWS.—The authority under subparagraph (A) of an employee, contractor, or detailee to submit to the Board a complaint or information shall be in addition to any other authority under another provision of law to submit a complaint or information. Any action taken under any other provision of law by the recipient of a complaint or information shall not preclude the Board from taking action relating to the same complaint or information.

(D) RELATIONSHIP TO ACTIONS TAKEN UNDER OTHER LAWS.—Nothing in this paragraph shall prevent—

(i) any individual from submitting a complaint or information to any authorized recipient of the complaint or information; or

(ii) the recipient of a complaint or information from taking independent action on the complaint or information."; and (B) by adding at the end the following: "(n) DEFINITIONS.—In this section, the terms 'congressional intelligence committees' and 'intelligence community' have the meanings given such terms in section 3 of the National Security Act of 1947 (50 U.S.C. 3003)." (2) PROHIBITED PERSONNEL PRACTICES.—Section 2302(b)(8)(B) of title 5, United States Code, is amended, in the matter preceding clause (i), by striking "or to the Inspector of an agency or another employee designated by the head of the agency to receive such disclosures" and inserting "the Inspector General of an agency, a supervisor in the employee's direct chain of command (up to and including the head of the employing agency), the Privacy and Civil Liberties Oversight Board, or an employee designated by any of the aforementioned individuals for the purpose of receiving such disclosures."

(c) PRIVACY AND CIVIL LIBERTIES OVERSIGHT BOARD SUBPOENA POWER.—Section 1061(g) of the Intelligence Reform and Terrorism Prevention Act of 2004 (42 U.S.C. 2000ee(g)) is amended—

(1) in paragraph (1)(D), by striking "submit a written request to the Attorney General of the United States that the Attorney General";

(2) by striking paragraph (2); and

(3) by redesignating paragraphs (3) and (4) as paragraphs (2) and (3), respectively.

(d) APPOINTMENT OF STAFF OF THE PRIVACY AND CIVIL LIBERTIES OVERSIGHT BOARD.—Section 1061(j) of the Intelligence Reform and Terrorism Prevention Act of 2004 (42 U.S.C. 2000ee(j)) is amended—

(1) by redesignating paragraphs (2) and (3) as paragraphs (3) and (4), respectively; and

(2) by inserting after paragraph (1) the following:

"(2) APPOINTMENT IN ABSENCE OF CHAIRMAN.—If the position of chairman of the Board is vacant, during the period of the vacancy the Board, at the direction of the majority of the members of the Board, may exercise the authority of the chairman under paragraph (1)."

(e) TENURE AND COMPENSATION OF PRIVACY AND CIVIL LIBERTIES OVERSIGHT BOARD MEMBERS AND STAFF.—

(1) IN GENERAL.—Section 1061 of the Intelligence Reform and Terrorism Prevention Act of 2004 (42 U.S.C. 2000ee), as amended by subsections (a) and (b), is further amended—

(A) in subsection (h)—

(i) in paragraph (1), by inserting "full-time" after "4 additional"; and

(ii) in paragraph (4)(B), by striking "except that" and all that follows through the end and inserting a period;

(B) (B) in subsection (i)(1)—

(i) in subparagraph (A), by striking "level III of the Executive Schedule under section 5314" and inserting "level II of the Executive Schedule under section 5313"; and

(ii) in subparagraph (B), by striking "level IV of the Executive Schedule" and all that follows through the end and inserting "level III of the Executive Schedule under section 5314 of title 5, United States Code."; and

(C) in subsection (j)(1), by striking "level V of the Executive Schedule under section 5316" and inserting "level IV of the Executive Schedule under section 5315."

(2) EFFECTIVE DATE; APPLICABILITY.—

(A) IN GENERAL.—The amendments made by paragraph (1)—

(i) shall take effect on the date of the enactment of this Act; and

(ii) except as provided in paragraph (2), shall apply to any appointment to a position as a member of the Privacy and Civil Liberties Oversight Board made on or after the date of the enactment of this Act.

(B) EXCEPTIONS.—

(i) COMPENSATION CHANGES.—The amendments made by subparagraphs (B)(i) and (C) of paragraph (1) shall take effect on the first day of the first pay period beginning after the date of the enactment of this Act.

(ii) ELECTION TO SERVE FULL TIME BY INCUMBENTS.— (I) IN GENERAL.—An individual serving as a member of the Privacy and Civil Liberties Oversight Board on the date of the enactment of this Act, including a member continuing to serve as a member under section 1061(h)(4)(B) of the Intelligence Reform and Terrorism Prevention Act of 2004 (42 U.S.C. 2000ee(h)(4)(B)), (referred to in this clause as a "current member") may make an election to—

(aa) serve as a member of the Privacy and Civil Liberties Oversight Board on a full-time basis and in accordance with section 1061 of the Intelligence Reform and Terrorism Prevention Act of 2004 (42 U.S.C. 2000ee), as amended by this section; or

(bb) serve as a member of the Privacy and Civil Liberties Oversight Board on a part-time basis in accordance with such section 1061, as in effect on the day before the date of the enactment of this Act, including the limitation on service after the expiration of the term of the member under subsection (h)(4)(B) of such section, as in effect on the day before the date of the enactment of this Act.

(II) ELECTION TO SERVE FULL TIME.—A current member making an election under subclause (I)(aa) shall begin serving as a member of the Privacy and Civil Liberties Oversight Board on a full- time basis on the first day of the first pay period beginning not less than 60 days after the date on which the current member makes such election.

(f) PROVISION OF INFORMATION ABOUT GOVERNMENT ACTIVITIES UNDER THE FOREIGN INTELLIGENCE SURVEILLANCE ACT OF 1978 TO THE PRIVACY AND CIVIL LIBERTIES OVERSIGHT BOARD.—The Attorney General shall fully inform the Privacy and Civil Liberties Oversight Board about any activities carried out by the Government under the Foreign Intelligence Surveillance Act of 1978 (50 U.S.C. 1801 et seq.), including by providing to the Board—

(1) copies of each detailed report submitted to a committee of Congress under such Act; and

(2) copies of each decision, order, and opinion of the Foreign Intelligence Surveillance Court or the Foreign Intelligence Surveillance Court of Review required to be included in the report under section 601(a) of such Act (50 U.S.C. 1871(a)).

Sec. 8. Improved Role in Oversight of Electronic Surveillance by Amici Curiae Appointed by Courts under Foreign Intelligence Surveillance Act of 1978.

(a) ROLE OF AMICI CURIAE GENERALLY.—

(1) IN GENERAL.—Section 103(i)(1) of the Foreign Intelligence Surveillance Act of 1978 (50 U.S.C. 1803(i)(1)) is amended by adding at the end the following: "Any amicus curiae designated pursuant to this paragraph may raise any issue with the Court at any time."

(2) REFERRAL OF CASES FOR REVIEW.—Section 103(i) of such Act is amended—

(A) by redesignating paragraphs (5) through (10) as paragraphs (6) through (11), respectively; and

(B) by inserting after paragraph (4) the following:

"(5) REFERRAL FOR REVIEW.—

"(A) REFERRAL TO FOREIGN INTELLIGENCE SURVEILLANCE COURT EN BANC.—If the court established under subsection (a) appoints an amicus curiae under paragraph (2)(A) to assist the Court in the consideration of any matter presented to the Court under this Act and the Court makes a decision with respect to such matter, the Court, in response to an application by the amicus curiae or any other individual designated under paragraph (1), may refer the decision to the Court en banc for review as the Court considers appropriate.

"(B) REFERRAL TO FOREIGN INTELLIGENCE SURVEILLANCE COURT OF REVIEW.—If the court established under subsection (a) appoints an amicus curiae under paragraph (2)(A) to assist the Court in the consideration of any matter presented to the Court under this Act and the Court makes a decision with respect to such matter, the Court, in response to an application by the amicus curiae or any other individual designated under paragraph (1) may refer the decision to the court established under subsection (b) for review as the Court considers appropriate.

"(C) REFERRAL TO SUPREME COURT.—If the Court of Review appoints an amicus curiae under paragraph (2) to assist the Court of Review in the review of any matter presented to the Court of Review under this Act or a question of law that may affect resolution of a matter in controversy and the Court of Review makes a decision with respect to such matter or question of law, the Court of Review, in response to an application by the amicus curiae or any other individual designated under paragraph (1) may refer the decision to the

Supreme Court for review as the Court of Review considers appropriate.

"(D) ANNUAL REPORT.—Not later than 60 days after the end of each calendar year, the Court and the Court of Review shall each publish, on their respective websites, a report listing—

(i) the number of applications for referral received by the Court or the Court of Review, as applicable, during the most recently concluded calendar year; and

(ii) the number of such applications for referral that were granted by the Court or the Court of Review, as applicable, during such calendar year."

(3) ASSISTANCE.—Section 103(i)(6) of such Act, as redesignated, is further amended to read as follows:

"(6) ASSISTANCE.—Any individual designated pursuant to paragraph (1) may raise a legal or technical issue or any other issue with the Court or the Court of Review at any time. If an amicus curiae is appointed under paragraph (2)(A)—

(A) the court shall notify all other amicus curiae designated under paragraph (1) of such appointment;

(B) the appointed amicus curiae may request, either directly or through the court, the assistance of the other amici curiae designated under paragraph (1); and

(C) all amici curiae designated under paragraph (1) may provide input to the court whether or not such input was formally requested by the court or the appointed amicus curiae."

(4) ACCESS TO INFORMATION.—Section 103(i)(7) of such Act, as redesignated, is further amended—

(A) in subparagraph (A)—

(i) in clause (i)—

(I) by striking "that the court" and inserting the following: "that—

"(I) the court"; and

(II) by striking "and" at the end and inserting the following: "or

"(II) are cited by the Government in an application or case with respect to which an amicus curiae is assisting a court under this subsection;";

(ii) by redesignating clause (ii) as clause (iii); and

(iii) by inserting after clause (i) the following:

"(ii) shall have access to an unredacted copy of each decision made by a court established under subsection (a) or (b) in which the court decides a question of law, notwithstanding whether the decision is classified; and";

(B) in subparagraph (B), by striking "may" and inserting "shall"; and

(C) in subparagraph (C)—

(i) in the subparagraph heading, by striking "CLASSIFIED INFORMATION" and inserting "ACCESS TO INFORMATION"; and

(ii) by striking "court may have access" and inserting the following: "court—

(i) shall have access to unredacted copies of each opinion, order, transcript, pleading, or other document of the Court and the Court of Review; and

(ii) may have access."

(5) PUBLIC NOTICE AND RECEIPT OF BRIEFS FROM THIRD PARTIES.—Section 103(i) of such Act, as amended by this subsection, is further amended by adding at the end the following:

"(12) PUBLIC NOTICE AND RECEIPT OF BRIEFS FROM THIRD PARTIES.—Whenever a court established under subsection (a) or (b) considers a novel question of law that can be considered without disclosing classified information, sources, or methods, the court shall, to the greatest extent practicable, consider such question in an open manner—

(A) by publishing on its website each question of law that the court is considering; and

(B) by accepting briefs from third parties relating to the question under consideration by the court."

(b) PARTICIPATION OF AMICI CURIAE IN OVERSIGHT OF AUTHORIZATIONS FOR TARGETING OF CERTAIN PERSONS OUTSIDE THE UNITED STATES OTHER THAN UNITED STATES PERSONS.—

(1) IN GENERAL.—Section 702(i)(2) of such Act (50 U.S.C. 1881a(i)(2)) is amended—

(A) in subparagraph (B), by redesignating clauses (i) and (ii) as subclauses (I) and (II), respectively, and adjusting the indentation of the margin of such subclauses, as so redesignated, two ems to the right;

(B) by redesignating subparagraphs (A) through (C) as clauses (i) through (iii), respectively, and adjusting the indentation of the margin of such clauses, as so redesignated, two ems to the right;

(C) by inserting before clause (i), as redesignated by subparagraph (B), the following:

(A) IN GENERAL.—"; and

(D) by adding at the end the following:

(B) PARTICIPATION BY AMICI CURIAE.—In reviewing a certification under subparagraph (A)(i), the Court shall randomly select an amicus curiae designated under section 103(i) to assist with such review."

(2) SCHEDULE.—Section 702(i)(5)(A) of such Act is amended by striking "at least 30 days prior to the expiration of such authorization" and inserting "such number of days before the expiration of such authorization as the Court considers necessary to comply with the requirements of paragraph (2)(B) or 30 days, whichever is greater."

(c) PUBLIC NOTICE OF QUESTIONS OF LAW CERTIFIED FOR REVIEW.—Section 103(j) of such Act (50 U.S.C. 1803(j)) is amended—

(1) by striking "Following" and inserting the following:

"(1) IN GENERAL.—Following"; and

(2) by adding at the end the following:

"(2) PUBLIC NOTICE.—

"(A) IN GENERAL.—Except as provided in subparagraph (B), whenever a court established under subsection (a) certifies a question of law for review under paragraph (1) of this subsection, the court shall publish on its website—

(i) a notice of the question of law to be reviewed; and

(ii) briefs submitted by the parties, which may be redacted at the discretion of the court to protect sources, methods, and other classified information.

"(B) PROTECTION OF CLASSIFIED INFORMATION, SOURCES, AND METHODS.—Subparagraph (A) shall apply to the greatest extent practicable, consistent with otherwise applicable law on the protection of classified information, sources, and methods."

Sec. 9. Reforms to the Foreign Intelligence Surveillance Court

(a) FISA COURT JUDGES.—

(1) NUMBER AND DESIGNATION OF JUDGES.—Section 103(a)(1) of the Foreign Intelligence Surveillance Act of 1978 (50 U.S.C. 1803(a)(1)) is amended to read as follows:

"(1)(A) There is a court which shall have jurisdiction to hear applications for and to grant orders approving electronic surveillance anywhere within the United States under the procedures set forth in this Act.

"(B)(i) The court established under subparagraph (A) shall consist of 13 judges, one of whom shall be designated from each judicial circuit (including the United States Court of Appeals for the District of Columbia and the United States Court of Appeals for the Federal Circuit).

"(ii) The Chief Justice of the United States shall—

(I) designate each judge of the court established under subparagraph (A) from the nominations made under subparagraph (C); and

(II) make the name of each judge of such court available to the public.

"(C)(i) When a vacancy occurs in the position of a judge of the court established under subparagraph (A) from a judicial circuit, the chief judge of the circuit shall propose a district judge for a judicial district within the judicial circuit to be designated for that position.

"(ii) If the Chief Justice does not designate a district judge proposed under clause (i), the chief judge shall propose 2 other district judges for a judicial district within the judicial circuit to be designated for that position and the Chief Justice shall designate 1 such district judge to that position.

(D) No judge of the court established under subparagraph (A) (except when sitting en banc under paragraph (2)) shall hear the same application for electronic surveillance under this Act which has been denied previously by another judge of such court.

(E) If any judge of the court established under subparagraph (A) denies an application for an order authorizing electronic surveillance under this Act, such judge shall provide immediately for the record a written statement of each reason for the judge's decision and, on motion of the United States, the record shall be transmitted, under seal, to the court of review established in subsection (b)."

(2) TENURE.—Section 103(d) of such Act is amended by striking "redesignation," and all that follows through the end and inserting "redesignation."

(3) IMPLEMENTATION.—

(A) INCUMBENTS.—A district judge designated to serve on the court established under subsection (a) of such section before the date of enactment of this Act may continue to serve in that position until the end of the term of the district judge under subsection (d) of such section, as in effect on the day before the date of the enactment of this Act.

(B) INITIAL APPOINTMENT AND TERM.—Notwithstanding any provision of such section, as amended by paragraphs (1) and (2), and not later than 180 days after the date of enactment of this Act, the Chief Justice of the United States shall—

(i) designate a district court judge who is serving in a judicial district within the District of Columbia circuit and proposed by the chief judge of such circuit to be a judge of the court established under section 103(a) of the Foreign Intelligence Surveillance Act of 1978 (50 U.S.C. 1803(a)) for an initial term of 7 years; and

(ii) designate a district court judge who is serving in a judicial district within the Federal circuit and proposed by the chief judge of such circuit to be a judge of such court for an initial term of 4 years.

(b) COURT OF REVIEW.—Section 103(b) of such Act is amended—

(1) by striking "The Chief Justice" and inserting "(1) Subject to paragraph (2), the Chief Justice"; and

(2) by adding at the end the following:

"(2) The Chief Justice may designate a district court judge or circuit court judge to a position on the court established under paragraph (1) only if at least 5 associate justices approve the designation of such individual."

Sec. 10. Study and Report on Diversity and Representation on the FISA Court and the FISA Court of Review

(a) STUDY.—The Committee on Intercircuit Assignments of the Judicial Conference of the United States shall conduct a study on how to ensure judges are appointed to the court established under subsection (a) of section 103 of the Foreign Intelligence Surveillance Act of 1978 (50 U.S.C. 1803) and the court established under subsection (b) of such section in a manner that ensures such courts are diverse and representative.

(b) REPORT.—Not later than 1 year after the date of the enactment of this Act, the Committee on Intercircuit Assignments shall submit to Congress a report on the study carried out under subsection (a).

Sec. 11. Grounds for Determining Injury in Fact in Civil Action Relating to Surveillance under Certain Provisions of Foreign Intelligence Surveillance Act of 1978

Section 702 of the Foreign Intelligence Surveillance Act of 1978 (50 U.S.C. 1881a), as amended by sections 2, 3, 4, 5, and 8(b), is further amended by adding at the end the following:

"(m) CHALLENGES TO GOVERNMENT SURVEILLANCE.—

"(1) INJURY IN FACT.—In any claim in a civil action brought in a court of the United States relating to surveillance conducted under this section, the person asserting the claim has suffered an injury in fact if the person—

(A) has a reasonable basis to believe that the person's communications will be acquired under this section; and

(B) has taken objectively reasonable steps to avoid surveillance under this section.

"(2) REASONABLE BASIS.—A person shall be presumed to have demonstrated a reasonable basis to believe that the communications of the person will be acquired under this section if the profession of the person requires the person regularly to communicate foreign intelligence information with persons who—

(A) are not United States persons; and

(B) are located outside the United States.

"(3) OBJECTIVE STEPS.—A person shall be presumed to have taken objectively reasonable steps to avoid surveillance under this section if the person demonstrates that the steps were taken in reasonable response to rules of professional conduct or analogous professional rules."

Sec. 12. Clarification of Applicability of Requirement to Declassify Significant Decisions of Foreign Intelligence Surveillance Court and Foreign Intelligence Surveillance Court of Review

Section 602 of the Foreign Intelligence Surveillance Act of 1978 (50 U.S.C. 1872) shall apply with respect to decisions, orders, and opinions described in subsection (a) of such section that were issued on, before, or after the date of the enactment of the Uniting and Strengthening America by Fulfilling Rights and Ensuring Effective Discipline Over Monitoring Act of 2015 (Public Law 114– 23).

Sec. 13. Clarification Regarding Treatment of Information Acquired under Foreign Intelligence Surveillance Act of 1978

(a) DERIVED DEFINED.—

(1) IN GENERAL.—Section 101 of the Foreign Intelligence Surveillance Act of 1978 (50 U.S.C. 1801) is amended by adding at the end the following:

"(q) For the purposes of notification provisions of this Act, information or evidence is 'derived' from an electronic surveillance, physical search, use of a pen register or trap and trace device, production of tangible things, or acquisition under this Act when the Government would not have originally possessed the information or evidence but for that electronic surveillance, physical search, use of a pen register or trap and trace device, production of tangible things, or acquisition, and regardless of any claim that the information or evidence is attenuated from the surveillance or search, would inevitably have been discovered, or was subsequently re-obtained through other means."

(2) POLICIES AND GUIDANCE.—

(A) IN GENERAL.—Not later than 90 days after the date of the enactment of this Act, the Attorney General and the Director of National Intelligence shall publish the following:

(i) Policies concerning the application of subsection (q) of section 101 of such Act, as added by paragraph (1).

(ii) Guidance for all members of the intelligence community (as defined in section 3 of the National Security Act of 1947 (50 U.S.C. 3003)) and all Federal agencies with law enforcement responsibilities concerning the application of such subsection.

(B) MODIFICATIONS.—Whenever the Attorney General and the Director modify a policy or guidance published under subparagraph (A), the Attorney General and the Director shall publish such modifications.

(b) USE OF INFORMATION ACQUIRED UNDER TITLE VII.—Section 706 of such Act (50 U.S.C. 1881e) is amended—

(1) in subsection (a), by striking "except for the purposes of subsection (j) of such section"; and

(2) by amending subsection (b) to read as follows:

"(b) INFORMATION ACQUIRED UNDER SECTIONS 703–705.—Information acquired from an acquisition conducted under section 703, 704, or 705 shall be deemed to be information acquired from an electronic surveillance pursuant to title I for the purposes of section 106."

Sec. 14. Limitation on Technical Assistance from Electronic Communication Service Providers under the Foreign Intelligence Surveillance Act of 1978

Section 702(h)(1) of the Foreign Intelligence Surveillance Act of 1978 (50 U.S.C. 1881a(h)(1)) is amended—

(1) by redesignating subparagraphs (A) and (B) as clauses (i) and (ii), respectively, and moving such clauses 2 ems to the right;

(2) by striking "With respect to" and inserting the following: "(A) IN GENERAL.—Subject to subparagraph (B), in carrying out"; and

(3) by adding at the end the following:

"(B) LIMITATIONS.—The Attorney General or the Director of National Intelligence may not request assistance from an electronic communication service provider under subparagraph (A) without demonstrating, to the satisfaction of the Court, that the assistance sought—

(i) is necessary;

(ii) is narrowly tailored to the surveillance at issue; and

(iii) would not pose an undue burden on the electronic communication service provider or its customers who are not an intended target of the surveillance.

"(C) COMPLIANCE.—An electronic communication service provider is not obligated to comply with a directive to provide assistance under this paragraph unless—

(i) such assistance is a manner or method that has been explicitly approved by the Court; and

(ii) the Court issues an order, which has been delivered to the provider, explicitly describing the assistance to be furnished by the provider that has been approved by the Court."

Sec. 15. Modification of Authorities for Public Reporting by Persons Subject to Nondisclosure Requirement Accompanying Order Under Foreign Intelligence Surveillance Act of 1978

(a) MODIFICATION OF AGGREGATION BANDING.—Subsection (a) of section 604 of the Foreign Intelligence Surveillance Act of 1978 (50 U.S.C. 1874) is amended—

(1) by striking paragraphs (1) through (3) and inserting the following:

"(1) A semiannual report that aggregates the number of orders, directives, or national security letters with which the person was required to comply into separate categories of—

"(A) the number of national security letters received, reported—

(i) for the first 1000 national security letters received, in bands of 200 starting with 1–200; and

(ii) for more than 1000 national security letters received, the precise number of national security letters received;

"(B) the number of customer selectors targeted by national security letters, reported—

(i) for the first 1000 customer selectors targeted, in bands of 200 starting with 1–200; and

(ii) for more than 1000 customer selectors targeted, the precise number of customer selectors targeted;

"(C) the number of orders or directives received, combined, under this Act for contents—

"(i) reported—(I) for the first 1000 orders and directives received, in bands of 200 starting with 1–200; and (II) for more than 1000 orders and directives received, the precise number of orders received; and

"(ii) disaggregated by whether the order or directive was issued under section 105, 402, 501, 702, 703, or 704;

"(D) the number of customer selectors targeted under orders or directives received, combined, under this Act for contents—

"(i) reported—(I) for the first 1000 customer selectors targeted, in bands of 200 starting with 1–200; and (II) for more than 1000 customer selectors targeted, the precise number of customer selectors targeted; and

"(ii) disaggregated by whether the order or directive was issued under section 105, 402, 501, 702, 703, or 704;

"(E) the number of orders or directives received under this Act for noncontents—

"(i) reported—(I) for the first 1000 orders or directives received, in bands of 200 starting with 1–200; and (II) for more than 1000 orders or directives received, the precise number of orders received; and

"(ii) disaggregated by whether the order or directive was issued under section 105, 402, 501, 702, 703, or 704; and

"(F) the number of customer selectors targeted under orders or directives under this Act for noncontents—

"(i) reported—(I) for the first 1000 customer selectors targeted, in bands of 200 starting with 1–200; and (II) for more than 1000 customer selectors targeted, the precise number of customer selectors targeted; and

"(ii) disaggregated by whether the order or directive was issued under section 105, 402, 501, 702, 703, or 704."; and

(2) by redesignating paragraph (4) as paragraph (2).

(b) ADDITIONAL DISCLOSURES.—Such section is amended—

(1) by redesignating subsections (b) through (d) as subsections (c) through (e), respectively; and

(2) by inserting after subsection (a) the following:

"(b) ADDITIONAL DISCLOSURES.—A person who publicly reports information under subsection (a) may also publicly report the following information, relating to the previous 180 days, using a semiannual report that indicates whether the person was or was not required to comply with an order, directive, or national security letter issued under each of sections 105, 402, 501, 702, 703, and 704 and the provisions listed in section 603(e)(3)."

Sec. 16. Annual Publication of Statistics on Number of Persons Targeted outside the United States under Certain Foreign Intelligence Surveillance Act of 1978 Authority

Not less frequently than once each year, the Director of National Intelligence shall publish the following:

(1) A description of the subject matter of each of the certifications provided under subsection (g) of section 702 of the Foreign Intelligence Surveillance Act of 1978 (50 U.S.C. 1881a) in the last calendar year.

(2) Statistics revealing the number of persons targeted in the last calendar year under subsection (a) of such section, disaggregated by certification under which the person was targeted.

Sec. 17. Repeal of Nonapplicability to Federal Bureau of Investigation of Certain Reporting Requirements under Foreign Intelligence Surveillance Act of 1978

Section 603(d)(2) of the Foreign Intelligence Surveillance Act of 1978 (50 U.S.C. 1873(d)(2)) is amended by striking "(A) FEDERAL BUREAU" and all that follows through "Paragraph (3)(B) of'' and inserting "Paragraph (3)(B)."

Sec. 18. Publication of Estimates Regarding Communications Collected under Certain Provision of Foreign Intelligence Surveillance Act of 1978

(a) IN GENERAL.—Except as provided in subsection (b), not later than 90 days after the date of the enactment of this Act, the Director of National Intelligence shall publish an estimate of—
  (1) the number of United States persons whose communications are collected under section 702 of the Foreign Intelligence Surveillance Act of 1978 (50 U.S.C. 1881a); or
  (2) the number of communications collected under such section to which a party is a person inside the United States.

(b) IN CASE OF TECHNICAL IMPOSSIBILITY.—If the Director determines that publishing an estimate pursuant to subsection (a) is not technically possible—
  (1) subsection (a) shall not apply; and
  (2) the Director shall publish an assessment in unclassified form explaining such determination, but may submit a classified annex to the appropriate committees of Congress as necessary.

(c) APPROPRIATE COMMITTEES OF CONGRESS DEFINED.—In this section, the term "appropriate committees of Congress" means—
  (1) the congressional intelligence committees (as defined in section 3 of the National Security Act of 1947 (50 U.S.C. 3003));
  (2) the Committee on the Judiciary of the Senate; and
  (3) the Committee on the Judiciary of the House of Representatives.

Sec. 19. Four-Year Extension of FISA Amendments Act of 2008

(a) EXTENSION.—Section 403(b) of the FISA Amendments Act of 2008 (Public Law 110–261) is amended—

(1) in paragraph (1) (50 U.S.C. 1881–1881g note), by striking "December 31, 2017" and inserting "September 30, 2021"; and

(2) in paragraph (2) (18 U.S.C. 2511 note), in the material preceding subparagraph (A), by striking "December 31, 2017" and inserting "September 30, 2021."

(b) CONFORMING AMENDMENT.—The heading of section 404(b)(1) of the FISA Amendments Act of 2008 (Public Law 110–261; 50 U.S.C. 1801 note) is amended by striking "DECEMBER 31, 2017" and inserting "SEPTEMBER 30, 2021."

In: DNA
Editor: Tomáš Koláček
ISBN: 978-1-53616-117-5

Chapter 4

DNA EVIDENCE: DOJ SHOULD IMPROVE PERFORMANCE MEASUREMENT AND PROPERLY DESIGN CONTROLS FOR NATIONWIDE GRANT PROGRAM*

United States Government Accountability Office

ABBREVIATIONS

AAG	Assistant Attorney General
ASCLD	American Society of Crime Laboratory Directors
BJA	Bureau of Justice Assistance
BJS	Bureau of Justice Statistics
DAEO	Designated Agency Ethics Official
DANY	New York County District Attorney's Office
DNA	deoxyribonucleic acid
DOJ	Department of Justice

* This is an edited, reformatted and augmented version of United States Government Accountability Office; Report to Congressional Requestors, Publication No. GAO-19-216, dated March 2019.

EI&CE	Forensic DNA Laboratory Efficiency Improvement and Capacity Enhancement grant program
CEBR	DNA Capacity Enhancement and Backlog Reduction grant program
CODIS	Combined DNA Index System
FBI	Federal Bureau of Investigation
IACP	International Association of Chiefs of Police
JMD	Justice Management Division
NIJ	National Institute of Justice
OGE	U.S. Office of Government Ethics
OMB	Office of Management and Budget
OJP	Office of Justice Programs
SAFE-ITR	Sexual Assault Forensic Evidence-Inventory, Tracking, and Reporting grant program
SAK	Sexual Assault Kit
SAKI	National Sexual Assault Kit Initiative

WHY GAO DID THIS STUDY

Many state and local government crime labs continue to have backlogs of requests for crime scene DNA analysis, which may include sexual assault kits, despite DOJ awarding nearly $1 billion since 2004 through the CEBR grant program to enhance lab capacity and reduce backlogs. Additionally, questions have been raised about potential improper connections among those who profit from CEBR grants—such as private labs and DNA equipment vendors— and those who advocate for CEBR funding.

This chapter examines, among other things, (1) what is known about the amount of backlogged DNA evidence at state and local government labs; (2) the extent to which OJP measures CEBR grant performance; and (3) the extent to which OJP has designed controls to identify conflicts of interest related to CEBR grants. GAO reviewed CEBR grantee data on DNA evidence backlogs from 2011-2017 (the most recent years of comparable data). GAO reviewed documentation on CEBR performance and controls related to conflicts of interest, transparency in grantee procurement, and lobbying. GAO also reviewed relevant reports and studies, and interviewed officials from DOJ.

What GAO Recommends

GAO is making four recommendations. OJP should (1) consistently document CEBR goals, (2) revise CEBR performance measures, (3) document its designation of confidential financial disclosure certifiers, and (4) clarify guidance for lobbying requirements for CEBR grantees. DOJ concurred with all four recommendations.

What GAO Found

GAO found that the reported number of backlogged requests for crime scene DNA analysis at state and local government labs has increased by 85 percent from 2011 through 2017, the most recent full year for which grantee data were available (from about 91,000 to about 169,000). This growth has occurred despite labs' collectively processing more requests over time, as shown below.

GAO found that the Office of Justice Programs (OJP)—the primary grant-making arm of the Department of Justice—has not consistently documented program-wide goals for its DNA Capacity Enhancement and Backlog Reduction grant program (CEBR). For example, it has documented different meanings of "capacity enhancement." Additionally, CEBR performance measures do not fully reflect selected attributes of successful performance measures, such as having measurable targets. These issues hinder OJP's ability to assess program results.

GAO found that OJP has designed controls to achieve its objectives related to compliance with selected requirements for transparency in grantee procurement that apply to both OJP and CEBR grantees.

However, OJP has not properly designed all controls related to selected requirements for conflicts of interest and lobbying. While OJP has designed a control to review confidential financial disclosure reports submitted by OJP employees who administer CEBR grants, OJP does not have the required documentation designating which officials are authorized to certify these reports. Similarly, OJP has designed controls to obtain required lobbying certification and disclosure forms from grantees, but OJP has not taken steps to ensure that grantees collect and forward to OJP, as appropriate, these forms from subgrantees and contractors, as required.

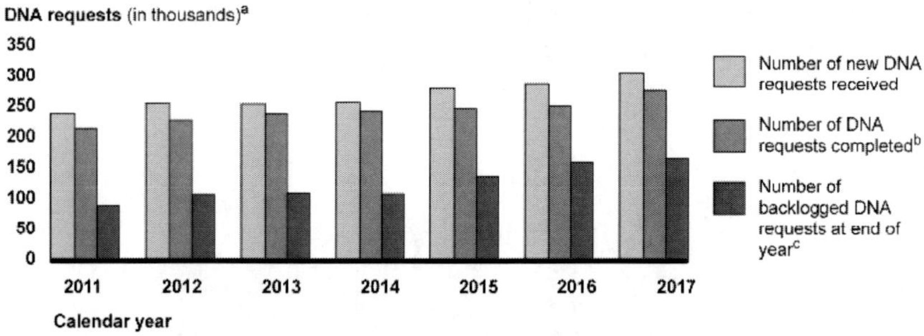

Source: GAO analysis of DNA Capacity Enhancement and Backlog Reduction (CEBR) grant program data. | GAO-19-216.

[a] "DNA requests" means requests for biology screening and/or DNA testing such that if a request requires one or both, it is counted once.

[b] Some requests may be closed by administrative means rather than through analysis, such as when a suspect pleads guilty before the evidence is analyzed or when the victim has not consented to participate in the criminal justice process. These requests are included in the number of requests received but are not included in the number of requests completed.

[c] The National Institute of Justice (NIJ) defines a "backlogged" request for crime scene DNA analysis as a request that has not been closed by the issuance of a report to the submitting agency within 30 days of receipt in the lab.

Requests for Crime Scene DNA Analysis, Requests Completed, and Backlogged Requests at State and Local Government Labs (2011-2017).

March 21, 2019

The Honorable Lindsey O. Graham
Chairman
Committee on the Judiciary
United States Senate

The Honorable Charles E. Grassley
United States Senate

At the end of 2017, about 169,000 requests for DNA analysis of crime scene evidence were backlogged at state and local government crime labs.[1] Since 2004, the Department of Justice (DOJ) has awarded nearly $1 billion to states and local jurisdictions through the DNA Capacity Enhancement and Backlog Reduction (CEBR) grant program (or CEBR legacy programs) to help increase lab capacity and reduce the amount of DNA evidence awaiting analysis at labs.[2] Nevertheless, although labs are processing more requests, backlogs of crime scene evidence in labs continue to increase. Further, CEBR grantees can use funds to contract with third parties—such as private labs and DNA equipment vendors—and questions have been raised about potential improper connections among these third parties and those who advocate for CEBR funding.

Additionally, thousands of sexual assault kits (SAKs) in law enforcement custody have not been submitted to crime labs for DNA analysis.[3] DOJ, state and local governments, and a nonprofit organization have efforts underway to inventory or quantify these "unsubmitted" SAKs. However, the number of unsubmitted SAKs nationwide is not well

---

[1] DNA stands for deoxyribonucleic acid and is the genetic material present in the nucleus of cells. According to the Department of Justice, a backlogged request for analysis of crime scene evidence is a request that has not been completed within 30 days of receipt in the laboratory.

[2] CEBR funds can be used for efforts related to processing (1) DNA evidence collected from crime scenes, and (2) DNA samples taken from convicted offenders, arrestees, and others pursuant to law.

[3] A SAK is a package of materials used to collect samples (evidence) from the victim's or suspect's body by a medical professional. The type of evidence collected depends on what occurred during the assault.

understood and there are concerns that these unsubmitted SAKs have delayed justice or enabled serial offenders to reoffend.

In 2013, we issued a report that reviewed, among other things, data and outcomes associated with DOJ's CEBR program (formerly named the DNA Backlog Reduction Program).[4] You asked us to revisit CEBR program data and outcomes and address additional questions related to backlogs of DNA evidence in labs and unsubmitted SAKs in law enforcement custody. In July 2018, we issued a testimony statement with preliminary observations from our ongoing work.[5]

This chapter addresses the following questions:

1) What is known about the amount of backlogged crime scene DNA evidence, including SAKs, in state and local government labs and the factors that contribute to such backlogs?
2) What is known about the amount of unsubmitted DNA evidence, including SAKs, in law enforcement custody and the factors that contribute to this unsubmitted evidence?
3) To what extent does the Office of Justice Programs (OJP)—the primary grant-making arm of DOJ—measure CEBR program performance?
4) To what extent has OJP designed controls related to conflicts of interest, transparency in grantee procurement, and lobbying requirements applicable to CEBR grants?

To describe what is known about the amount of backlogged crime scene DNA evidence in state and local labs, we collected yearly baseline data from the National Institute of Justice (NIJ), the component within OJP that administers the CEBR program, for calendar years 2011 through 2017.[6] We selected these years because in 2011 two previous grant

---

[4] GAO, *Justice Grant Programs: DOJ Could Improve Decision-Making Documentation and Better Assess Results of DNA Backlog Reduction Program Funds*, GAO-13-605 (Washington, D.C.: July 2013).

[5] GAO, *DNA Evidence: Preliminary Observations on DOJ's DNA Capacity Enhancement and Backlog Reduction Program*, GAO-18-651T (Washington, D.C. July 2018).

[6] NIJ collects baseline data—which includes data from all labs within grantees' jurisdictions (not only those labs that use CEBR grant funds)—as part of the grant application process. We

programs were combined into what is now known as the CEBR grant program, and because 2017 is the most recent full calendar year for which CEBR grantee data were available at the time of our review. To assess the extent to which NIJ-compiled CEBR baseline data are reliable, we completed a number of data reliability steps, including discussing data entry issues with grantees and running logic tests on the data, among others. We determined that the CEBR baseline data are sufficiently reliable for our purposes.[7] To ensure we had identified all possible sources of data and to better understand how labs collect and report data, we met with knowledgeable officials from various entities, including national associations, research groups, CEBR grantees or grantee labs, and DOJ.

To identify and describe what is known about the amount of unsubmitted DNA evidence, including SAKs, in law enforcement custody, we obtained and reviewed information and/or data related to several efforts to inventory or quantify unsubmitted SAKs in law enforcement custody, including:

- Data from DOJ's National Sexual Assault Kit Initiative (SAKI) grant program;
- Data from the New York County District Attorney's Office (DANY) SAK grant program;
- State laws requiring SAK inventories, including state-specific data resulting from such laws in Idaho and Texas;[8]
- Information from reporting requirements under the Sexual Assault Forensic Evidence Reporting (SAFER) Act;

---

also reviewed data from other sources but determined that CEBR data were the most useful for showing aggregate nationwide trends. For our rationale for using CEBR data (as opposed to other sources), see the full description of our scope and methodology in appendix I. For more information on these other sources of data, see appendix II.

[7] We determined that state and local government labs participating in the CEBR program from 2011 through 2017 generally represent the level of workloads and backlogs from state and local government labs that participate in CODIS nationwide. However, we found that CEBR turnaround time data for 2017 were not reported as consistently as data for workloads and backlogs and may not be representative of nationwide turnaround times for state and local government labs.

[8] We selected Idaho and Texas as illustrative examples because, relative to each other, they differed with respect to legal requirements for conducting inventories, and timeframes associated with conducting those inventories.

- Data from the Joyful Heart Foundation, a national nonprofit survivors' advocacy organization; and
- Data from two prior nationwide studies on the amount of, and reasons for, unsubmitted DNA evidence in law enforcement custody (including DNA evidence associated with various types of crime, not just sexual assault).

We include general information about all of these efforts and data from some of these efforts in this chapter. For the data we include in this chapter (SAKI, DANY, Idaho and Texas), we present all available data for the time periods the efforts have been active. We assessed the reliability of the data from these efforts and found they were sufficiently reliable for our purposes. For details about the specific steps we took to assess the reliability of data for each of these efforts, see appendix I.

To identify and describe factors that contribute to backlogs of unanalyzed DNA evidence at laboratories and unsubmitted DNA evidence, including SAKs, in law enforcement custody, we reviewed 22 reports, including 16 government reports (or government-funded reports), 4 academic journal articles, 1 book, and 1 study from a non-governmental organization. Five of the government (or government-funded) reports included nationwide studies of public crime labs and law enforcement agencies with original research. We also discussed these factors with a non-generalizable selection of DNA evidence stakeholders from 17 entities, including DOJ officials, CEBR grantees, forensic crime lab directors, and relevant academics and practitioners. For information about how we selected the reports we reviewed and the stakeholders we interviewed, see appendix I. Finally, we conducted legal research on state laws that require law enforcement to submit SAKs for analysis and that also require labs to analyze previously-unanalyzed SAKs. We summarized information from these sources to identify common factors, and we included illustrative examples of the types of factors we identified in this chapter.

To evaluate how OJP measures CEBR program performance, we first sought to identify CEBR program-wide goals. To do this, we reviewed OJP CEBR documentation, including the most recent CEBR grant solicitation and NIJ reports that include CEBR performance information. We also discussed CEBR goals with NIJ officials. We then assessed these CEBR program-wide goals against federal internal control standards that call for management to define goals clearly.[9] As part of this effort, we also obtained and reviewed information on OJP's 6 CEBR program-wide performance measures, and evaluated the extent to which they reflected attributes of successful performance measures we previously established.[10]

To evaluate the extent to which OJP has designed controls related to conflicts of interest, transparency in grantee procurement, and lobbying requirements applicable to CEBR grants, we conducted legal research on these topic areas to identify federal statutes and regulations applicable to OJP and CEBR applicants, recipients, and subrecipients.[11] To determine the extent to which OJP has designed controls to achieve its objectives related to compliance with these requirements, we reviewed DOJ or OJP documentation, which included, among others, the Fiscal Year 2018 CEBR solicitation, the DOJ Departmental Ethics Website, and the DOJ Ethics Handbook for On and Off-Duty Conduct. To further determine the extent to which OJP has designed each control, we also obtained and reviewed a variety of CEBR program documents, including those associated with reviewing and approving CEBR grant applications, and grant monitoring. We also discussed some of the controls that we identified with four CEBR

---

[9] GAO, *Standards for Internal Control in the Federal Government*, GAO-14-704G (Washington, D.C.: September 2014), Principle 6 – Define Objectives and Risk Tolerances (starts on page 35).

[10] Successful performance measures help assess progress toward preestablished goals. Our prior work established nine attributes of successful performance measures: linkage, clarity, measurable targets, objectivity, reliability, core program activities, limited overlap, balance, and government-wide priorities. GAO, *Tax Administration: IRS Needs to Further Refine Its Tax Filing Season Performance Measures*, GAO-03-143, (Washington, D.C.: November 2002). We selected 3 of the 9 attributes (linkage, clarity, and measurable targets) because they are foundational. By foundational, we mean that without them, other attributes are less relevant or important. We selected 2 of the 9 attributes (core program activities, balance) because they assess the extent to which the performance measures cover a variety of aspects of performance. For additional detail about our selection of these criteria, see appendix I.

[11] We did not review the extent to which controls were implemented effectively (e.g., through controls testing procedures such as conducting case file reviews), nor did we review the extent to which employees were trained to implement controls effectively (e.g., through reviewing training material and/or outcomes of such training). We examined requirements applicable to subgrantees and contractors under CEBR grantees and subgrantees, and subcontractors.

grantees. Finally, we assessed controls that were not properly designed against federal internal control standards, which specify how management should design controls to achieve its compliance objectives.[12]

We conducted this performance audit from September 2017 to March 2019 in accordance with generally accepted government auditing standards. Those standards require that we plan and perform the audit to obtain sufficient, appropriate evidence to provide a reasonable basis for our findings and conclusions based on our audit objectives. We believe that the evidence obtained provides a reasonable basis for our findings and conclusions based on our audit objectives.

BACKGROUND

Using DNA Evidence from Crime Scenes to Aid Investigations

DNA evidence from crime scenes can be used to help link perpetrators or victims to each other and to crime scenes, and ultimately bring perpetrators to justice. Figure 1 illustrates stages in the process of using DNA evidence from crime scenes to aid investigations.

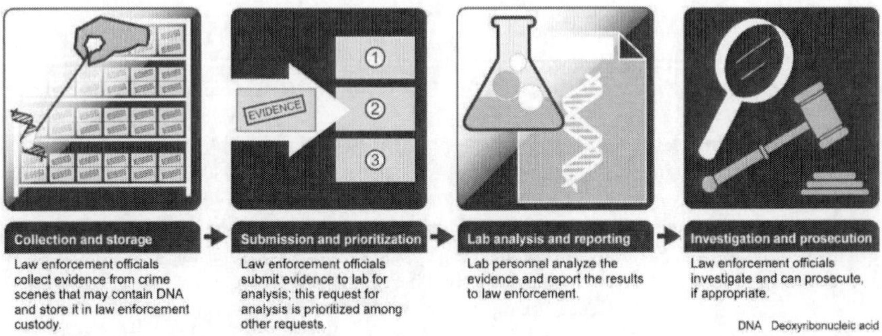

Source: GAO analysis. | GAO-19-216.

Figure 1. Stages in Using DNA Evidence from Crime Scenes to Aid Investigations.

---

[12] GAO-14-704G. Per guidance in these standards, an existing control is not properly designed when, even if the control operates as designed, the control objective would not be met.

Collection and Storage

After an alleged crime, law enforcement officials collect evidence that may contain DNA from crime scenes, victims, and suspects (hereafter referred to as "DNA evidence"). In the case of sexual assault, a victim may choose to undergo a medical-forensic exam where a medical professional collects samples from the victim's body.[13] This evidence is placed in a SAK for storage and for potential future DNA analysis. According to NIJ, the contents of a SAK vary by jurisdiction but generally include swabs and collection envelopes for biological materials, debris, and underwear.

Most states have established statutes and/or policies for biological evidence retention, according to a report supported by the National Institute of Standards and Technology (NIST) and NIJ.[14] Biological evidence refers to samples of biological material—such as hair, tissue, bones, teeth, blood, semen, or other bodily fluids—or to evidence items containing biological materials. However, this chapter states that existing state laws vary in (1) their definitions of what constitutes biological evidence for purposes of evidence retention, (2) categories of crime for which biological evidence should be retained, and (3) the length of time biological evidence should be retained.

Submission and Prioritization

Law enforcement officials and prosecutors may submit requests for analysis of DNA evidence to labs.[15] These officials may opt not to submit

---

[13] In this report, we use the term "sexual assault" to include attacks or attempted attacks generally involving unwanted sexual contact between victim and offender, including rape.

[14] In 2013, the National Institute of Standards and Technology—a component of the Department of Commerce—and NIJ jointly supported the issuance of a report detailing best practices for biological evidence preservation. See Technical Working Group on Biological Evidence Preservation, *The Biological Evidence Preservation handbook: Best Practices for Evidence Handlers*, a report jointly supported NIST and NIJ, April 2013.

[15] Other groups may also submit requests, such as public defenders and those who submit requests on behalf of convicted offenders (seeking to clear the innocent of wrongful convictions). According to NIJ and stakeholders we spoke with, all requests are not equal and do not take the same amount of resources. Specifically, requests for analysis may contain one or more "items" for examination, such as weapons, carpets, or bedsheets. Each item may contain multiple "samples" for analysis (e.g., multiple stains on clothing). Each sample, in turn, may be subject to multiple "tests," such as instrumental analysis, extractions, or comparative examinations. Thus, depending on the number of items,

DNA evidence to labs if they do not believe the results of the analysis would add value to the case. However, laws in some states require officials to submit DNA evidence to labs for analysis in cases of sexual assault. Additionally, DOJ, the International Association of Chiefs of Police (IACP), and the National Sheriffs' Association, have reported that all SAKs should be analyzed if victims report the crime to law enforcement. According to OJP, the majority of sexual assaults are committed by persons known to the victim.[16] Although analysis of SAKs in these cases may not provide new information about the alleged perpetrator's identity, the results may link the alleged perpetrator to other crimes within and across jurisdictions, as we discuss later. According to officials from NIJ, IACP, and the American Society of Crime Lab Directors (ASCLD), law enforcement officials and prosecutors who submit requests for DNA analysis, and labs that process requests, generally prioritize requests based on various factors. These factors may include, among others, state laws, public safety concerns, the trial date of the associated case, whether there is a known suspect, or the type of crime—such as homicide, sexual assault, or property crime. According to these officials, law enforcement officials or prosecutors also generally have the option to mark a request as needing to be "expedited" or "rushed," based on the same factors.

Lab Analysis and Reporting

Once submitted and prioritized, labs perform analysis on DNA evidence. We define "DNA analysis" in this chapter as (1) biology screening (locating, screening, identifying, and characterizing blood and other biological stains and substances); and/or (2) DNA testing (identifying and comparing DNA profiles in biological samples).[17] A DNA profile is generally considered a unique genetic identifier based on the genetic

---

samples, and tests needed, requests can vary greatly in the amount of resources needed to conduct the analysis.

[16] According to OJP's Bureau of Justice Statistics (BJS), from 1994 through 2010, about 75 to 80 percent of rapes and sexual assaults were perpetrated by someone known to the victim. DOJ, OJP, BJS, *Female Victims of Sexual Violence, 1994-2010*, (March 2013).

[17] We use this definition of "DNA analysis" because the CEBR program counts requests that undergo biology screening and/or DNA testing as one request (such that if a request requires one or both, it is counted once).

constitution of an individual.[18] Comparing DNA profiles from crime scene evidence to DNA profiles from known sources can help lab analysts identify the source of unknown DNA profiles found on crime scene evidence. Thus, DNA samples used for such comparisons, called "reference" samples, are generally collected from victims and may also be collected from others whose DNA may be present in crime scene evidence—such as suspects, crime scene personnel, first responders, and consensual sexual partners (in cases of sexual assault).[19]

Upon completing DNA analysis and required reviews, the lab provides a report to law enforcement with the results.[20] The results detail the extent to which any DNA profiles identified from crime scene evidence match DNA profiles from reference samples, and note whether any profiles that may be attributable to a potential perpetrator are eligible to be uploaded into the FBI's Combined DNA Index System (CODIS), among other things. CODIS is a system that allows federal, state, and local labs to exchange and compare DNA profiles electronically in order to develop investigative leads. CODIS has multiple levels where DNA profiles can be stored and searched: the local level, the state level, and the national level.[21]

---

[18] A DNA profile contains the genetic constitution of an individual at defined locations (also known as loci) in the DNA. Each person (except identical twins) has a unique DNA profile when used in the context of the national level of the FBI's Combined DNA Index System (CODIS), which now evaluates 20 specific DNA locations. We discuss CODIS later in this report.

[19] Some sources distinguish between "reference" samples (from victims and suspects), and "elimination" samples (from others whose DNA may be present in crime scene evidence). However, for purposes of this report, we refer to all these samples as "reference" samples. In practice, evidence often contains a mixture of DNA from more than one person. These mixtures can be challenging to analyze and interpret. Additionally, DNA analysis may result in "partial" profiles, which may occur when samples have low quantities of DNA or are exposed to extreme environmental conditions, among other things. Not all biological evidence that undergoes DNA analysis yields usable DNA.

[20] Labs must perform technical and administrative reviews prior to reporting results to law enforcement, according to the FBI's Quality Assurance Standards (QAS), which labs that participate in the national level of CODIS must adhere to. The technical review is an evaluation of reports, notes, data, and other documents to ensure there is an appropriate and sufficient basis for the scientific conclusions. The administrative review is an evaluation of the report and supporting documentation for consistency with laboratory policies and for editorial correctness.

[21] According to FBI officials, the FBI sets requirements for what DNA profiles are eligible to be uploaded into the national level of CODIS; and state and local governments set requirements for what DNA profiles are eligible to be uploaded at state and local levels of CODIS, respectively.

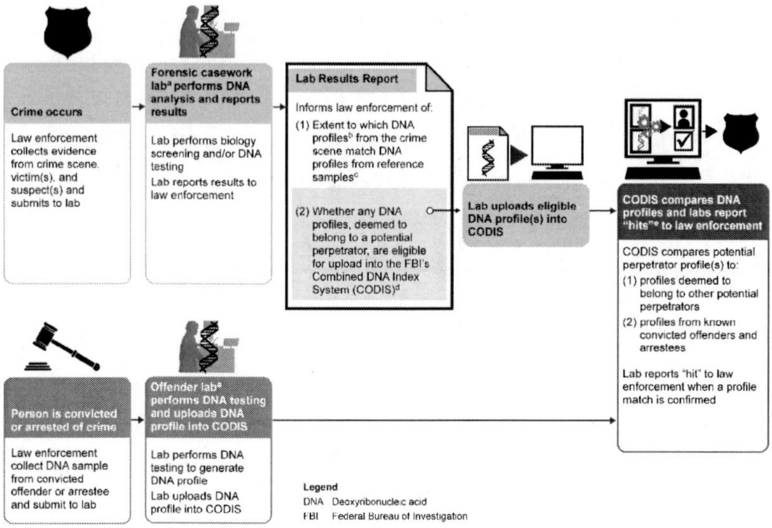

Source: GAO analysis of DOJ and stakeholder information; Art Explosion (clip art). | GAO-19-216.

[a] We use the term "forensic casework lab" to refer to labs that perform DNA analysis on evidence from crime scenes, victims, and suspects; we use the term "offender lab" to refer to labs that perform DNA analysis on samples taken from convicted offenders, arrestees, and other categories of persons as authorized by law; some labs do both.

[b] A DNA profile is generally considered to be a unique genetic identifier based on the genetic constitution of an individual.

[c] "Reference samples," as used in this chapter, are DNA samples collected from known individuals— such as victims, suspects, crime scene personnel, first responders, and consensual sexual partners—for purposes of comparison to recovered crime scene DNA.

[d] CODIS is the generic term used to describe the FBI's program of support for criminal justice DNA databases as well as the software used to run these databases. CODIS has multiple levels where DNA profiles can be stored and searched; the local level (for city and county DNA labs), state level, and national level. According to FBI officials, the FBI sets requirements for what DNA profiles are eligible to be uploaded into CODIS at the national level; and state and local governments set requirements for what DNA profiles are eligible to be uploaded at state and local levels of CODIS, respectively.

[e] The FBI defines a hit as "a confirmed match that aids an investigation and one or more of the case(s) involved in the match are unsolved."

Figure 2. How DNA Profiles Are Generated and Compared to Help Law Enforcement.

Once labs upload one or more eligible DNA profiles into CODIS, CODIS compares the profile(s) to (1) profiles generated from evidence taken from other crime scenes and deemed to belong to other potential perpetrators; and (2) profiles generated from samples taken from convicted offenders and arrestees as authorized by law.[22] When a profile deemed to belong to a potential perpetrator matches another profile within CODIS, a "hit" or investigative lead may be developed.[23] A hit may involve the linking of crimes to each other and/or to convicted offenders or arrestees. If this occurs, the lab will send a second report to law enforcement notifying them of the hit and providing additional information. Figure 2 below shows how CODIS can help develop investigative leads.

Investigation and Prosecution

Only federal, state, or local government labs that meet the FBI's Quality Assurance Standards can participate in CODIS.[24] Government labs may outsource DNA analysis or testing to private labs, but government labs must take ownership of the results of those analyses pursuant to FBI

[22] According to DOJ, the federal government, all 50 states, the District of Columbia, and Puerto Rico have laws requiring the collection of DNA samples from individuals convicted of certain crimes; in addition, according to DOJ, the federal government and over half the states have laws authorizing the collection of DNA from individuals arrested for certain crimes. According to FBI officials, federal statute authorizes the collection DNA samples from non-U.S. persons who are detained under the authority of the United States, and other categories of persons, as authorized by either federal or state law. 34 U.S.C. § 40702.

[23] The FBI defines a hit as, "A confirmed match that aids an investigation and one or more of the case(s) involved in the match are unsolved." According to FBI officials, not all profile matches result in "hits" because the matches may relate to cases already solved or may not provide new information. For instance, a match may confirm the identity of a perpetrator whose identity had already been confirmed by DNA analysis prior to entry into CODIS.

[24] FBI, *Quality Assurance Standards for Forensic DNA Testing Laboratories* (September 1, 2011). FBI, *Quality Assurance Standards for DNA Databasing Laboratories* (September 1, 2011). According to FBI officials, the integrity of CODIS relies on the permanent ownership and maintenance of the samples that produced DNA profiles, and its participating labs provide such permanence. See also, 34 U.S.C. § 12592 (providing that only information on DNA identification records and analyses that are maintained by federal, state, or local criminal justice agencies (or the Secretary of Defense) can be included in CODIS).

Quality Assurance Standards before entering any associated DNA data into CODIS.[25]

Upon receiving the results of lab analysis or CODIS hit information from a lab, law enforcement officials are responsible for investigating cases further and, as appropriate, prosecuting potential perpetrators. Additional resources may be needed for investigation, prosecution, and victim notification when jurisdictions receive many CODIS hits in a short period of time, such as after a jurisdiction decides to analyze all its old SAKs.

"Unsubmitted" DNA Evidence and the Two Types of DNA Evidence "Backlogs"

"Unsubmitted" crime scene DNA evidence in law enforcement custody is separate from, but related to, "backlogged" crime scene DNA evidence in labs. Further, there are two separate types of DNA evidence backlogs: (1) backlogs of crime scene evidence in labs, and (2) backlogs of convicted offender and arrestee samples in labs (see Figure 3).

"Unsubmitted" Crime Scene DNA Evidence versus "Backlogged" Crime Scene DNA Evidence

Efforts, systems, and methods used to audit and inventory "unsubmitted" and "backlogged" crime scene DNA evidence are different.[26] NIJ uses the term "unsubmitted" to refer to DNA evidence that has not been submitted to labs but remains in law enforcement custody.[27]

---

[25] According to NIJ, taking 'ownership' of data implies and requires a technical review of all the data. Moreover, NIJ reported that the requirement to technically review all of the data is a major bottleneck in labs that sometimes mitigates the expediency of outsourcing.

[26] Although unsubmitted DNA evidence in law enforcement custody may be considered by some as part of a "backlog," we do not use this this term to describe unsubmitted evidence for purposes of clarity.

[27] For purposes of this report, "unsubmitted" SAKs also refer to SAKs being stored in laboratories, hospitals, or other medical facilities but for which a request for analysis has not been made. IACP officials said that most SAKs are stored in law enforcement storage facilities, but they may also be found in labs, hospitals, rape crisis centers, or other medical facilities.

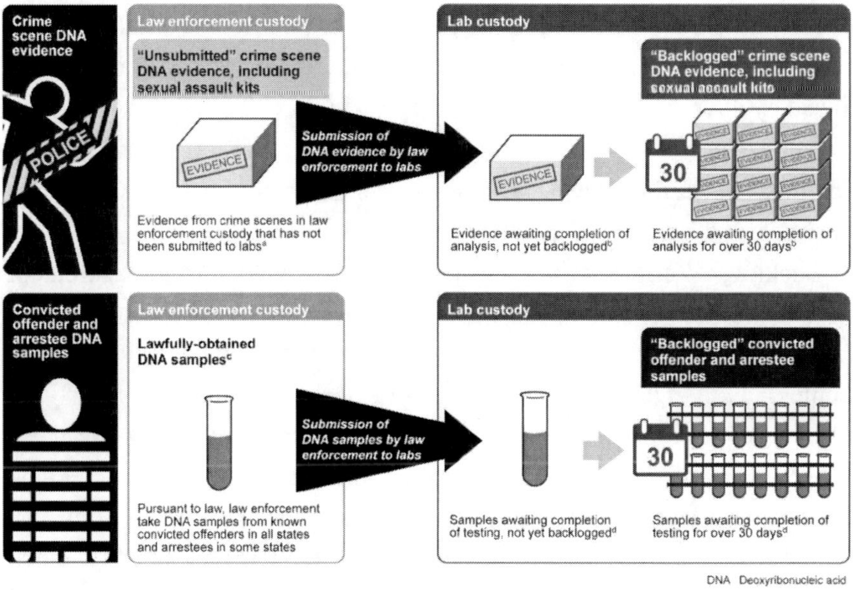

Source: GAO analysis (using definitions of backlogs from the National Institute of Justice). | GAO-19-216.

[a] Not all crime scene evidence collected by law enforcement is submitted to labs. Law enforcement may collect more evidence than is needed and labs may have capacity limitations.

[b] NIJ defines a "completed" request for analysis of crime scene evidence as a request that has been closed by a report to law enforcement; and NIJ defines a "backlogged" request as one that has not been completed within 30 days of receipt in the lab.

[c] According to OJP, DNA has not been collected from potentially thousands of convicted offenders across the United States due to the lack of coordinated DNA collections in some jurisdictions, as well as the frequent inability for rural and smaller municipalities to collect samples during processing.

[d] NIJ defines a "completed" DNA convicted offender or arrestee sample as a sample that has been tested and uploaded to the FBI's Combined DNA Index System; and NIJ defines a "backlogged" request one that has not been completed within 30 days of receipt in the lab.

Figure 3. "Unsubmitted" Crime Scene DNA Evidence and the Two Types of DNA Evidence "Backlogs."

In contrast, NIJ uses the term "backlog" to refer to a request for DNA analysis that has been submitted to a lab but that has not been completed (closed by a report to the submitting agency) within 30 days of receipt.[28]

Backlogs of Crime Scene Evidence versus Backlogs of Convicted Offender and Arrestee Samples

Backlogs of crime scene DNA evidence are separate from backlogs of convicted offender and arrestee DNA samples. DNA analysis of crime scene evidence and DNA testing of convicted offender and arrestee samples involve different processes, and some labs only perform one function or the other. According to NIJ, analyzing evidence from crime scenes is time consuming because it often must undergo biology screening to determine if biological materials are present before DNA testing can even begin. Further, some of the samples can be degraded or fragmented and can contain DNA from multiple contributors. In contrast, according to NIJ, DNA samples taken from convicted offenders and arrestees are easier, faster to analyze, and less expensive to analyze because they are taken from a single source (the convicted offender or arrestee) and generally do not require rigorous interpretation.

According to officials from NIJ, the FBI, and ASCLD, backlogs of convicted offender and arrestee samples are not a significant issue today. See appendix III for more information about backlogs of convicted offender and arrestee samples.

DOJ Grant Programs That Address DNA Evidence Backlogs at Labs and Unsubmitted SAKs in Law Enforcement Custody

DOJ has two grant programs that primarily address DNA evidence backlogs at state and local government labs and two grant programs that

---

[28] According to NIJ officials, "receipt" refers to receipt in the area of the lab that performs DNA analysis. Evidence associated with a given request may need to first be processed in one or more other areas of the lab—such as areas that perform latent fingerprint analysis or firearms analysis—before being processed by the area of the lab that performs DNA analysis.

primarily address unsubmitted SAKs in law enforcement custody. Specifically, the CEBR program and the Forensic DNA Laboratory Efficiency Improvement and Capacity Enhancement program (EI&CE) address DNA evidence backlogs at labs; and the Sexual Assault Kit Initiative (SAKI) and the Sexual Assault Forensic Evidence—Inventory, Tracking, and Reporting program (SAFE-ITR) address unsubmitted SAKs in law enforcement custody. We primarily focus on the CEBR program in this chapter but have included additional information on all these grant programs in appendix IV.

CEBR is a grant program that dates back to 2004.[29] Grant awards are made non-competitively to states and units of local government based on a formula set by DOJ that allocates certain amounts to each state.[30] This formula takes into account each state's population and associated crime levels, and guarantees a minimum amount for eligible applicants from each state. In fiscal year 2018 OJP awarded $67.8 million in CEBR grants to 127 grantees in all 50 states and the District of Columbia.

The CEBR program is funded by an appropriation "for a DNA analysis and capacity enhancement program and for other local, State, and Federal forensic activities."[31] The broad appropriations language enables NIJ to allocate this funding for a variety of forensic programs. However, the Justice for All Reauthorization Act of 2016 mandated that at least 75 percent of funds made available under this appropriation be used for grants

---

[29] In 2011, grant programs that separately funded labs that analyzed (1) crime scene DNA evidence, and (2) convicted offender and arrestee DNA samples, were combined into one grant program called the DNA Backlog Reduction Program. In 2014, this was renamed the DNA Capacity Enhancement and Backlog Reduction Program.

[30] The formula is set by DOJ administratively as opposed to some formula grant programs where the formula is set by statute.

[31] The appropriation language states that funds are "for a DNA analysis and capacity enhancement program and for other local, State, and Federal forensic activities, including the purposes authorized under section 2 of the DNA Analysis Backlog Elimination Act of 2000 (Public Law 106–546) (the Debbie Smith DNA Backlog Grant Program)." The purposes of the CEBR grant program are generally similar to the purposes of the Debbie Smith DNA Backlog Grant Program. There is no additional statutory authorization for the program.

for DNA analysis or to increase the capacity of government labs to carry out DNA analysis.[32]

Reported Backlog of Requests for Crime Scene DNA Analysis at State and Local Government Labs Nationwide Is Increasing due to Various Factors

Data from DOJ's CEBR grant program show that the number of backlogged requests for analysis of DNA evidence at state and local government labs nationwide has increased from 2011 through 2017. Specifically, labs have continued to receive more requests for DNA analysis than they can complete each year, even though labs have consistently completed an increasing number of requests over time. These data also show that average turnaround time for these requests has stayed relatively constant among CEBR grantees, though these data may not be representative of turnaround times for state and local government labs nationwide. We identified several factors that are reported to have contributed to this increased demand for DNA analysis, including scientific advancements in DNA analysis technology and state laws requiring analysis of certain DNA evidence, among other factors.

---

[32] The Justice for All Reauthorization Act of 2016, Pub. L. No. 114-324, § 3(a), requires that not less than 75 percent of the funds made available under this appropriation be provided for grants for activities described under paragraphs (1), (2), and (3) of section 2(a) of the DNA Analysis Backlog Elimination Act of 2000. Those purposes include (1) To carry out, for inclusion in the Combined DNA Index System of the Federal Bureau of Investigation, DNA analyses of samples collected under applicable legal authority; (2) To carry out, for inclusion in such Combined DNA Index System, DNA analyses of samples from crime scenes, including samples from rape kits, samples from other sexual assault evidence, and samples taken in cases without an identified suspect; and (3) To increase the capacity of laboratories owned by the State or by units of local government to carry out DNA analysis of samples specified in the purposes above. Prior to this, congressional reports accompanying the appropriation have directed that OJP make funding for DNA analysis and capacity enhancement a priority. OJP officials said they use CEBR funding to help meet the "75 percent" requirement discussed above.

CEBR Data Show an Increasing Nationwide Backlog of Requests for Crime Scene DNA Analysis at State and Local Government Labs

Among CEBR grantees, the reported aggregated backlog of requests for crime scene DNA analysis has increased by nearly 85 percent from 2011 through 2017 (from about 91,000 to about 169,000).[33] As of January 1, 2018 there were 194 state and local government labs in the United States that performed DNA analysis on evidence from crime scenes and participated in CODIS.[34]

These CEBR data show that from 2011 through 2017 requests for DNA analysis outpaced labs' ability to complete new and existing requests each year, resulting in a backlog. However, labs have consistently completed more requests over time, as shown in figure 4 below. According to our analysis, state and local government labs participating in the CEBR program from 2011 through 2017 generally represent the level of workloads and backlogs from state and local government labs that perform DNA analysis on evidence taken from crime scenes and participate in CODIS nationwide.[35]

---

[33] As part of the grant application process, NIJ requests applicants for CEBR grants to provide data from all labs in their jurisdiction, even if certain labs will not be using CEBR funds.

[34] As of January 1, 2018, there were 201 total government labs that participated in CODIS in the U.S. Of these 201, 3 are federal labs and are thus not eligible for CEBR grants; of the remaining 198 state and local government labs, 194 perform DNA analysis on evidence taken from crime scenes and 56 perform analysis on samples taken from convicted offenders and arrestees (52 labs do both, which is why the two numbers are greater than 198).

[35] NIJ seeks to collect data from CEBR grantees in years they did not apply for CEBR funding to keep the data consistent; we included this data in our analysis. According to information provided by OJP, as of January 1, 2018, CEBR grantees represented 189 of the 194 state and local government labs in the U.S. that perform DNA analysis on evidence taken from crime scenes and participate in CODIS. For the 5 labs not included, we determined that they constituted a small portion of backlog data in other years so as not to substantively affect results (for example, less than three percent since 2012). We were unable to identify the number of labs that did not participate in the CEBR program for years 2011 through 2016. However, NIJ officials said that CEBR grantees have generally been representative of all state and local government labs in the U.S. that participated in CODIS in prior years as well. In addition, to ensure that fluctuations in the population of CEBR grantees during these years did not significantly affect our results, we compared data on the number of requests backlogged and the annual rate of change in the number of requests backlogged for each year from all grantees (including those grantees who reported data from some but not

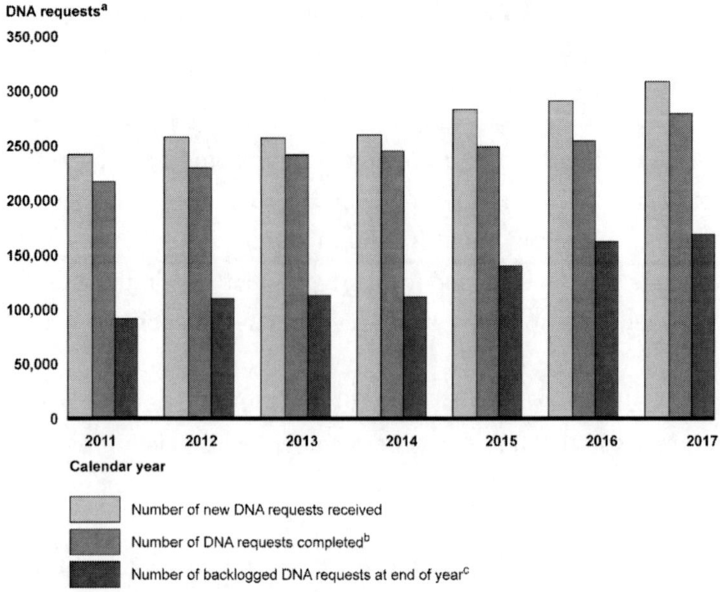

Source: GAO analysis of DNA Capacity Enhancement and Backlog Reduction (CEBR) grant program data. | GAO-19-216.

[a] By "DNA requests" we mean requests for biology screening and/or DNA testing such that if a request requires one or both, it is counted once.

[b] Some requests may be closed by administrative means rather than through analysis, such as when a suspect pleads guilty before the evidence is analyzed or when the victim has not consented to participate in the criminal justice process. These requests are included in the number of requests received but are not included in the number of requests completed.

[c] NIJ defines a "backlogged" request for crime scene DNA analysis as a request that has not been closed by the issuance of a report to the submitting agency within 30 days of receipt in the lab. The number of backlogged requests is larger than the difference between the number of new requests and the number of requests completed because backlogs also include outstanding requests from previous years.

Figure 4. Requests for Crime Scene DNA Analysis, Requests Completed, and Backlogged Requests at State and Local Government Labs, from Calendar Year 2011 through 2017.

all years between 2011 and 2016) to data from a subset of grantees (including those that provided data for all years between 2011 through 2016). We found that any fluctuations did not affect overall conclusions about trends in DNA analysis in these prior years.

In addition to reviewing CEBR data on workloads and backlogs, we also reviewed CEBR data to better understand average reported turnaround times. NIJ defines turnaround time for a request for DNA evidence as the time from when the lab receives the evidence to the time when the lab issues a report with the results of the analysis to the submitting agency. From 2011 through 2017, the average turnaround time for requests across CEBR grantees remained relatively constant, around 150 days.[36] Average turnaround time was generally higher for requests associated with nonviolent crimes than requests associated with violent crimes during this same time period. These turnaround time data provide insights but may not be representative of turnaround times for state and local government labs nationwide because the number of grantees that reported turnaround time data was smaller than those that reported workload and backlog data.[37]

Nevertheless, these data are consistent with what stakeholders we interviewed—including representatives from NIJ, IACP, and ASCLD—said about turnaround times. They said that labs generally prioritize requests associated with violent crime, such as homicide and sexual assault, ahead of those associated with less-violent crime, such as property crimes.[38] Regarding sexual assault specifically, representatives from ASCLD and IACP said law enforcement and labs generally prioritize requests where the victim did not know the identity of the alleged offender

---

[36] The averages we present in this paragraph represent a weighted mean of the reported turnaround time for requests for DNA analysis in each calendar year. To obtain a weighted average for each calendar year, we weighted the mean turnaround time reported by each grantee by the number of requests completed in the calendar year.

[37] CEBR grantees reported turnaround times less frequently than they reported other data, such as the number of requests received, the number of requests completed, and backlogged requests. For instance, for 2017 data, 40 of 138 grantees reported average turnaround times for requests associated with violent crimes, 42 of 138 grantees reported average turnaround times for requests associated with nonviolent crimes, and 89 of 138 grantees reported average overall turnaround times (turnaround times that include all requests). These aggregated average turnaround times may not fully reflect turnaround times associated with "expedited" or "rushed" requests. Specifically, when NIJ collects turnaround times from grantees, NIJ asks for turnaround times associated with non-priority (i.e., non-expedited or non-rushed) requests. However, according to ASCLD officials, crime labs' IT systems may or may not be able to separate requests into those that were expedited and those that were not expedited for purposes of reporting turnaround time to NIJ.

[38] According to IACP officials, prioritization practices may also be influenced by state law.

(i.e., stranger sexual assault) ahead of requests where the identity of the alleged offender is known.[39]

Data on Backlogged SAKs at Labs

The CEBR program does not specifically collect data on the number of backlogged SAKs at labs. NIJ officials said the CEBR program is designed to support DNA analysis for all crimes, thus NIJ does not focus on obtaining data associated with only one type of crime (i.e., sexual assault).

According to NIJ officials, obtaining data on the number of SAKs awaiting analysis may be burdensome for labs or may produce unreliable data. According to NIJ and ASCLD officials, SAKs are considered one piece of evidence within a request that may contain other pieces of evidence, such as weapons, carpets, and bedsheets—and labs' IT systems may not be configured to track each piece of evidence separately. However, according to an ASCLD official and one lab official we interviewed, some labs have modified their IT systems to be able to track SAKs, usually in response to state laws mandating the tracking or reporting of SAKs.

Additionally, DNA stakeholders used differing terminology to identify SAKs within their labs and IT systems, which may further complicate reporting on the number of SAKs awaiting analysis at labs. Specifically, some DNA stakeholders described a SAK as an "item" with multiple "samples" inside; others described a SAK as an "item" with multiple "items" or "sub-items" inside; and yet others disagreed that SAKs were considered an "item" altogether and preferred the term "piece of evidence" or "package of items."

Source: GAO analysis of NIJ and stakeholder interviews.| GAO-19-216.

Although aggregate trends show an increase in the nationwide backlog, individual labs have varying levels of requests awaiting analysis and differ in average turnaround times. For example, among the 82 grantees for which we had complete data from 2011 through 2017, 22 grantees (27 percent) reported an overall decrease in the backlog during this time.

---

[39] The Joyful Heart Foundation, a survivors' advocacy organization, stated that as awareness around the sexual assault kit backlog and the value of DNA evidence has increased in the past several years, sexual assault is being more readily included in prioritization.

Additionally, while the average reported turnaround time (including requests associated with both violent and nonviolent crime) in 2017 was about 150 days, averages among grantees that spanned from 20 to 580 days.

NIJ does not collect data on the number of backlogged SAKs at labs and NIJ officials stated that it may be difficult to quantify these data.

In addition to CEBR grantee data, we identified two other sources of nationwide data on DNA evidence processing at government labs: (1) the Bureau of Justice Statistics (BJS) Census of Publicly Funded Forensic Crime Labs, and (2) Project FORESIGHT, a research project at West Virginia University's College of Business and Economics. We determined that the most useful data for showing aggregate nationwide trends was CEBR data. Information about this determination can be found in appendix I. Information on these additional data sources and considerations for reviewing DNA evidence data can be found in appendix II.

Scientific Advancements and Other Factors Have Increased Demand for DNA Analysis of Crime Scene Evidence beyond Labs' Capacities

Based on a review of a selection of literature and discussions with DNA evidence stakeholders, we identified the following factors that are reported to have contributed to an increased demand for DNA analysis of crime scene evidence beyond labs' capacities, thereby contributing to increased backlogs:[40]

- Scientific advancements. Recent scientific advancements have increased the quality of DNA analysis by allowing lab analysts to obtain DNA profiles from smaller amounts of biological evidence. For instance, "touch DNA" samples—which become available when DNA is transferred by the simple touching of an object—can

[40] For more information about how we selected the literature and a list of the stakeholders we interviewed for this review, see appendix I.

now be used to yield DNA profiles. Further, many older and unsolved cases from the "pre-DNA" era are being reopened and subjected to DNA analysis in hopes of solving these crimes. Thus, scientific advancements have increased the amount of evidence that is eligible to be analyzed and, as a result, have increased the demand for DNA analysis.

- Decreases in turnaround times. Labs that decrease their turnaround time for DNA analysis may see corresponding increases in requests from law enforcement, according to preliminary results from Project FORESIGHT, a research project designed to help crime labs evaluate their efficiency.[41] Other DNA stakeholders, including NIJ and labs that we met with also made similar observations. Thus, as law enforcement officials are able to obtain the results of DNA analysis from labs more quickly, they may request DNA analysis more often.
- Increased awareness among law enforcement. Increased awareness among law enforcement officers of the value of DNA analysis in solving current and older cases has led to law enforcement agencies submitting more DNA evidence to labs for analysis. Additionally, law enforcement officials' awareness of the usefulness of CODIS has contributed to increased demand. Specifically, NIJ and other stakeholders we interviewed stated that the volume of DNA profiles in CODIS from convicted offenders and arrestees as well as potential perpetrators has increased significantly over recent years. This, in turn, has increased the likelihood of obtaining a "hit" or investigative lead when comparing a DNA profile from a potential perpetrator to DNA profiles in CODIS.[42]

---

[41] Project FORESIGHT, run out of West Virginia University's College of Business and Economics, is a research project designed to help crime labs evaluate their efficiency and inform resource allocation decisions. Project FORESIGHT obtains data from public and private labs that participate in the project.

[42] Additionally, when deciding whether to submit DNA evidence for analysis, law enforcement and prosecutors may consider jurors' expectations that DNA analysis is presented. The literature and stakeholders commonly refer to this as the "CSI effect," which is named after a television program that features the use of DNA in solving criminal cases.

- Recent legislation requiring SAK analysis. State legislation requiring SAK analysis has contributed to an increase in demand for DNA analysis. As of September 2018, we identified at least 27 states that had enacted laws requiring law enforcement to submit SAKs for testing that come into law enforcement possession.[43] Eleven of these states also required the submission for testing of previously untested SAKs. Twenty-three of these laws were passed in 2014 or later.

In addition to the factors that have contributed to increased demand, resource challenges and constraints on government labs' capacities are reported to have helped contribute to backlogs. State and local labs generally receive appropriations from state or local governments, which may direct local funding priorities based on state law or local policy. Additionally, some of these labs report facing lengthy hiring and training processes for forensic analysts, and often lose staff to private or federal labs which may offer higher pay, further limiting lab capacity for completing analysis. Officials at one lab stated that increased quality standards from the FBI and accrediting bodies also require lab resources. Further, according to NIJ, some labs face space constraints that limit their ability to hire additional staff or purchase additional equipment.

The Amount of Unsubmitted DNA Evidence Nationwide Is Unknown and Various Factors Contribute to This Unsubmitted Evidence

According to our review of data sources and the DNA evidence stakeholders with whom we spoke, there have not been any recent attempts to quantify levels of all types of unsubmitted DNA evidence in law enforcement custody nationwide. However, certain state and local

[43] Legislation varies by state. For instance, some states may require the analysis of SAKs as well as other evidence relating to sexual assault, or may require that unreported SAKs not be analyzed.

jurisdictions have performed inventories of their previously unsubmitted SAKs, some of which have been supported by DOJ's National Sexual Assault Kit Initiative (SAKI) grant program. We also identified factors that are reported to have contributed to unsubmitted SAKs, as well as other types of DNA evidence in law enforcement custody. These include archived biological evidence from the pre-DNA era and limited lab capacity, among other factors.

The Amount of Unsubmitted DNA Evidence Nationwide Is Unknown though Some Jurisdictions Have Inventoried Previously Unsubmitted SAKs

We were unable to find data that could definitively quantify levels of unsubmitted DNA evidence in law enforcement custody nationwide. According to DNA evidence stakeholders and our own review of potential data sources, there have been no recent attempts to quantify unsubmitted DNA evidence associated with all types of crime nationwide.

Regarding non-SAK DNA evidence, stakeholders said there may be reasons why there have been no recent attempts to quantify this unsubmitted evidence. Specifically, they stated that it is difficult to quantify non-SAK DNA evidence because this evidence is generally not confined to standardized containers or known to contain biological evidence. In contrast, SAK evidence is generally confined to a box which is thought to contain at least some biological evidence. Further, there has been strong advocacy for the analysis of SAKs, which some stakeholders said has influenced the passage of state laws requiring SAK inventories and analysis, and otherwise increased attention on unsubmitted SAKs.

Regarding SAKs specifically, although there is no definitive count of unsubmitted SAKs in law enforcement custody nationwide, certain state and local jurisdictions have performed inventories to identify previously unsubmitted SAKs. Some of these efforts have been supported by SAKI grants. The information below provides a description of three large-scale efforts to inventory and/or submit to labs for analysis previously unsubmitted SAKs in law enforcement custody. Additional information on these and other efforts are in appendix V, along with a list of challenges associated with inventorying SAKs within jurisdictions or quantifying them across jurisdictions.[44]

Sexual Assault Kit Initiative Grantee Data

SAKI grants provide funding to state law enforcement agencies and units of local government to help them address unsubmitted SAKs. SAKI grants require grantees to first take an inventory of unsubmitted SAKs; however, according to BJA officials, some jurisdictions had already identified unsubmitted SAKs prior to receiving SAKI funds. As of June 2018, 38 SAKI grantees identified a total of 102,837 previously unsubmitted SAKs (59,614 SAKs identified prior to receiving SAKI funds, and 43,223 identified after receiving SAKI funds).

New York County District Attorney's Office Grantee Data

In 2015, New York County District Attorney's Office (DANY) awarded nearly $38 million in grants to 32 jurisdictions (grantees) across 20 states to help them analyze SAKs. DANY and SAKI program administrators coordinated to ensure that the DANY and SAKI programs complemented each other and did not duplicate resources. Funding awarded through DANY's program is only allowed to be used for costs associated with analyzing SAKs; thus, DANY did not provide funds for grantees to perform inventories.[45] Nevertheless, DANY tracks the number of SAKs submitted to labs for analysis, which DANY officials said serves

---

[44] The additional information in appendix V includes data on activities resulting from inventorying and/or submitting for DNA analysis previously unsubmitted SAKs, including counts of SAKs for which DNA analysis was completed, and counts of resulting criminal investigations, among other things. Data from these efforts may overlap, therefore it is not possible to combine totals from each effort without potentially duplicating results.

[45] The DANY grant solicitation did not distinguish between "unsubmitted" SAKs in law enforcement jurisdiction and "backlogged" SAKs in laboratories. Specifically, according to the DANY SAK Program grant solicitation, a SAK eligible for analysis with grant funds is one that is connected to a reported sexual assault that has not been analyzed within 365 days of being booked into law enforcement evidence regardless of the reason it had not been analyzed.

as a proxy for previously unsubmitted SAKs. As of September 2018, DANY program grantees submitted 62,915 SAKs to labs for analysis.

Data from State Efforts to Inventory Unsubmitted SAKs

As of September 2018, we identified at least 26 states with laws that require the conducting of SAK inventories where evidence had not yet been submitted for lab analysis. We found that legislation varied by state. For example, some states require one-time inventories and other states require annual inventories; additionally, some states require inventories of SAKs and others require inventories of sexual assault cases (which often include SAKs). Twenty-four of these laws were passed in 2014 or later.

For example, Idaho passed a law requiring a one-time inventory of unsubmitted SAKs by December 2016 and an annual inventory of SAKs beginning in January 2017. As of September 2018, Idaho had identified 1,933 unsubmitted SAKs. Similarly, Texas passed a law requiring a one-time inventory of sexual assault cases with unsubmitted DNA evidence that was collected between September 1, 1996 and July 1, 2011. As of August 2017, Texas reported 18,955 cases.

Archived Evidence Not Subjected to DNA Analysis and Other Factors Contribute to Unsubmitted DNA Evidence in Law Enforcement Custody

Based on a review of a selection of literature and discussions with DNA evidence stakeholders, we identified the following factors that are reported to have contributed to unsubmitted DNA evidence—including non-SAK DNA evidence and SAKs—being stored in law enforcement custody. These factors are as follows:[46]

---

[46] For information about how we selected the literature and a list of the stakeholders we interviewed for this review, see Appendix I.

- Archived biological evidence not subjected to DNA analysis. For several decades before the relatively recent scientific advancements in DNA analysis and the widespread use of CODIS, law enforcement submitted evidence to labs to identify or analyze biological material to support investigations, and then archived that evidence for future use. For instance, they may have performed biology screening to identify what fluid was on clothing, or they may have tested blood samples to see if there was a blood type match. In addition, some evidence with biological material remains in law enforcement storage because other forensic evidence was used and the case was adjudicated without performing DNA analysis. For example, the lab could use other evidence such as a fingerprint to identify or confirm a potential perpetrator and the suspect could plead guilty before the lab analyzed the evidence.
- Evidence collection practices. Crime scene investigators often collect more evidence from crime scenes than is needed because they only have one chance to collect such evidence and they may not know which evidence will be most useful to the case. This evidence may or may not contain biological material and may remain in law enforcement custody, depending on the circumstances of the case and relevant laws and policies.
- Limited lab capacity. Labs may face resource constraints such as limited funding and lengthy hiring and training processes which can limit overall lab capacity for completing analysis. As a result, some labs may implement policies which limit the amount of DNA evidence that law enforcement officials can submit. In addition, stakeholders said some law enforcement agencies may not submit DNA evidence to a lab if the lab has previously been slow to provide results.
- Misunderstandings between law enforcement and labs about evidence submission policies. In some jurisdictions, law enforcement policy may determine what evidence law enforcement agencies submit to labs, while in other jurisdictions lab policy may

be the guiding protocol. Misunderstandings between law enforcement and labs about each other's policies can be a source of confusion. In some instances, stakeholders said this may cause investigators not to submit evidence to the lab. For example, a law enforcement officer may not submit evidence to the lab for analysis if he or she believes that the crime lab will be unwilling to accept it or unable to analyze it.

- Case-specific determinations for SAKs. Law enforcement officials may not submit DNA evidence if they do not believe the evidence will add value to the case.[47] For example, if a suspect in a sexual assault case claims that a sexual encounter was consensual, DNA analysis may not provide any value to the specific investigation because the suspect admitted to the sexual encounter. Further, law enforcement officials' perception of the victim's credibility or the circumstances of the case may influence the decision to pursue an investigation and submit DNA evidence for analysis. Law enforcement officials may have other reasons for not submitting SAKs, such as when charges have not been filed or the alleged offender has already been adjudicated.
- Sexual assault victim consent for DNA analysis. After undergoing a sexual assault exam at a hospital or rape crisis center, sexual assault victims may not want their SAKs submitted for DNA analysis for various reasons. NIJ refers to these SAKs—where the victim has consented to the collection of the SAK but has not consented to participate in the criminal justice process—as "unreported" SAKs. According to NIJ, an unreported SAK cannot be submitted to a laboratory for analysis, unless applicable law provides. As a result, the DNA evidence either remains in the hospital or rape crisis center, or is transferred to law enforcement where it remains in law enforcement custody.

---

[47] In many jurisdictions, law enforcement officials and prosecutors no longer have the ability to make case-specific determinations because state laws mandate SAK submission.

CEBR Program Goals Are Not Consistently Documented and Its Performance Measures Do Not Fully Reflect Selected Attributes of Successful Performance Measures

CEBR Program Goals Are Not Consistently Documented so as to Clearly Communicate Intended Program Results

We identified statements in NIJ and CEBR program documentation that communicated program-wide goals, but the documentation did not consistently identify the same goals or the same number of goals.[48] For example, a stated goal of improving the quality of DNA testing was included in 2 of 4 NIJ documents we reviewed and increasing the efficiency of testing was only stated in one of these documents. NIJ officials clarified that the CEBR program has two goals: (1) to increase laboratory capacity for DNA analysis, and (2) to reduce backlogs of DNA evidence awaiting analysis. The inconsistency of the goals can be seen across goal statements outlined in various NIJ sources as seen in Table 1 below.

While NIJ officials' explanations help clarify their goals for us, such clarifications are not documented and thus are not available to congressional decision-makers who help determine if the CEBR program is achieving its intended results. Moreover, after NIJ officials clarified that the CEBR program has a goal to reduce backlogs, as described in the last row of Table 1, NIJ later reported that eliminating the nationwide backlog is not a program goal. Officials stated they believe the goal of eliminating backlogs is unachievable in the foreseeable future because increases in demand for DNA analysis are driven by factors outside of NIJ's control. Thus, officials said they are not comfortable setting an unachievable goal and reporting data related to that goal that may be misinterpreted.

---

[48] Program-wide goals are different than goals for individual grants. The CEBR grant program allows grantees to create their own goals and objectives within the framework of overall program-wide goals.

Table 1. DNA Capacity Enhancement and Backlog Reduction Grant Program (CEBR) Goals Listed in National Institute of Justice (NIJ) Documentation and NIJ Officials' Clarifications

Source	CEBR Goal(s)
Goals as stated in NIJ documentation	
CEBR grant solicitation (fiscal year 2018)	The overarching goal is to assist eligible States and units of local government to increase laboratory capacity and reduce the number of samples awaiting analysis in both the casework and database sections.[a]
NIJ website (2018)[b]	The goals are to assist eligible states and units of local government to: • Process, record, screen and analyze forensic DNA and/or DNA database samples. • Increase the capacity of public forensic DNA and DNA database laboratories to process more DNA samples, thereby helping to reduce the number of forensic DNA and DNA database samples awaiting analysis.
NIJ report (2018)[c]	The main objectives are to improve the quality of testing and increase the efficiency of evidence processing in forensic DNA laboratories.
NIJ report (2017)[d]	The goal is to help increase the throughput of evidence processing at the nation's forensic DNA laboratories and improve the quality of testing.
NIJ officials' clarification of goals	
NIJ officials' clarification (2018)	The CEBR program has two goals: Increase laboratory capacity for DNA analysis. This refers to increasing samples analyzed, reducing processing times, and increasing the number of DNA profiles uploaded into the FBI's Combined DNA Index System (CODIS)—all while either maintaining or increasing the quality of DNA analysis at labs. Reduce backlogs of DNA evidence awaiting analysis. This refers to reducing the number of backlogged requests awaiting analysis by more than the number of requests that become backlogged during the same timeframe.

Source: GAO analysis of NIJ documentation and interview with NIJ officials. | GAO-19-216.

[a] "Sections" refers to areas within labs that perform different types of analysis; the "casework" section refers to the section that performs DNA analysis on evidence from crime scenes, victims, and suspects; the "database" section refers to the section that performs DNA analysis on samples collected from convicted offenders, arrestees, and others as authorized by law.

[b] DOJ, OJP, NIJ, DNA Capacity Enhancement and Backlog Reduction Program, https://www.nij.gov/topics/forensics/lab-operations/evidence-backlogs/Pages/backlog-reductionprogram.aspx. Accessed June 4, 2018.

[c] DOJ, OJP, NIJ, Report Forensic Science Fiscal Year 2017 Funding for DNA Analysis, Capacity Enhancement, and Other Forensic Activities. (Washington D.C.: April 2018). NIJ officials stated that this chapter is intended for a different audience than grant applicants to emphasize that improvements in quality and throughput are achieved through innovations such as more sensitive chemistries, faster technologies, and streamlined workflows, all of which, according to NIJ officials, can help increase efficiency and decrease backlogs.

[d] DOJ, OJP, NIJ, Report Forensic Science Fiscal Year 2016 Funding for DNA Analysis, Capacity Enhancement, and Other Forensic Activities. Washington D.C.: May 2017). NIJ officials stated that this chapter is intended for a different audience than grant applicants to emphasize that improvements in quality and throughput are achieved through innovations such as more sensitive chemistries, faster technologies, and streamlined workflows, all of which, according to NIJ officials, can help increase efficiency and decrease backlogs.

However, they also feel they need to have a goal to reduce or eliminate backlogs because this, they said, was an original purpose of the program. This further exacerbates the lack of consistency and clarity available to Congress.[49] *Standards for Internal Control in the Federal Government* state that management should define goals clearly to enable the identification of risks and define risk tolerances.[50] In doing so, management defines goals in clear and specific terms so they can be understood, including clearly defining what is to be achieved. Consistently documenting CEBR program goals would better position NIJ to clearly communicate intended program results.

CEBR Performance Measures Do Not Fully Reflect Selected Attributes of Successful Performance Measures

We found that the CEBR program's six program-wide performance measures do not fully reflect the five selected attributes of successful performance measures we evaluated them against.[51] The purpose of performance measures is to assess progress against goals, and while there is no single correct method to develop successful performance measures, we have selected attributes that, if met, will help performance measures assess progress toward goals.[52]

---

[49] NIJ officials reported that they have responded to Congress numerous times in the past and have provided context about the CEBR program and challenges related to backlog reduction.

[50] GAO, *Standards for Internal Control in the Federal Government*, GAO-14-704G (September 2014).

[51] NIJ requires data from grantees related to other performance measures, such as DNA analysis throughput for the lab and average turnaround times. However, we did not include these data or measures in our scope because NIJ does not use them for program-wide reporting.

[52] Our prior work establishes 9 attributes of successful performance measures: linkage, clarity, measurable targets, objectivity, reliability, core program activities, limited overlap, balance, and government-wide priorities. GAO, *Tax Administration: IRS Needs to Further Refine Its Tax Filing Season Performance Measures, GAO-03-143 (Washington, D.C.: November 2002)*. We selected 3 of the 9 attributes (linkage, clarity, and measurable targets) because they are foundational. By foundational, we mean that without them, other attributes are less relevant or important. We selected 2 of the 9 attributes (core program activities, balance) because they assess the extent to which the performance measures cover a variety of aspects of performance. For additional detail about our selection of these criteria, see appendix I.

Table 2. Assessment of DNA Capacity Enhancement and Backlog Reduction Program-wide Performance Measures against Five Selected Attributes of Successful Performance Measures

	Linkage Measure is aligned with program goals	**Clarity** Measure is clearly stated and is consistent with the methodology used to calculate it	**Measurable Target** Measure has a numerical goal	**Core Program Activities** As a group, measures cover the activities that an entity is expected to perform to support the intent of the program	**Balance** As a group, measures ensure that key program priorities are covered
Number of Forensic Cases Analyzed[a]	●	●	○	○ (Evaluated as a group)	◖ (Evaluated as a group)
Number of Forensic DNA Profiles[b] Uploaded to the FBI's Combined DNA Index System (CODIS)[c]	●	●	○		
Percentage Increase in Forensic DNA Profiles Uploaded to CODIS from the Previous Year	●	◖	●		
Number of Convicted Offender and/or Arrestee Database Samples Analyzed[d]	●	●	○		
Number of Convicted Offender and/or Arrestee Database Samples Uploaded to CODIS	●	●	○	○ (Evaluated as a group)	◖ (Evaluated as a group)
Number of CODIS Hits[e]	●	●	○		

Legend: Measure fully reflects criterion ● Measure partially reflects criterion ◖ Measure does not reflect criterion ○

Source: GAO analysis of CEBR program-wide performance measures. | GAO-19-216

Note: Gray shading represents attributes that are foundational to successful performance measurement systems. By "foundational," we mean that, without them, other attributes are less relevant or important.

Note: For all measures except one—percentage increase in forensic DNA profiles uploaded to CODIS from the previous year—NIJ directs grantees to only include requests analyzed or samples tested where CEBR funds were used to help with overtime, supplies, outsourcing, and/or personnel.

[a] "Forensic cases" refers to requests for DNA analysis on evidence from crime scenes, victims, and suspects.

[b] A DNA profile is generally considered a unique genetic identifier based on the genetic constitution of an individual.

[c] CODIS is a system that allows federal, state, and local labs to exchange and compare DNA profiles electronically. As a result of processing evidence from crime scenes, only DNA profiles believed to belong to a potential perpetrator can be uploaded into CODIS.

[d] "Convicted offender and/or arrestee database samples" refers to DNA analysis on samples taken from convicted offenders, arrestees, and other categories of persons as authorized by law.

[e] The FBI defines a hit as "a confirmed [DNA profile] match that aids an investigation and one or more of the case(s) involved in the match are unsolved."

See Table 2 for a summary of our analysis, and see appendixes VI and VII for more information about CEBR performance measures and the criteria we evaluated them against.

Linkage

We determined that all six performance measures aligned with at least one CEBR program goal, although the goals are not consistent across documentation, as we have discussed.

Clarity

We found that 5 of the 6 measures are clearly stated and the names and definitions are consistent with the methodologies used to calculate them.[1] We also found that one measure—the Percentage Increase in Forensic DNA Profiles Uploaded to CODIS from the Previous Year—is partially consistent with the methodology used to calculate it. Specifically, by referencing "profiles uploaded to CODIS," the name of the measure implies that it includes all DNA profiles uploaded into CODIS when, upon clarification, NIJ officials stated that it only reflects DNA profiles uploaded to the national level of CODIS.[2]

Additionally, we found that the name of the measure—as stated in OJP's Fiscal Year 2018 Performance Budget (the most recent OJP performance budget in which the measure was included)—contained two inconsistencies in the wording, as compared to the documentation NIJ provided us that clarifies how the measure is worded and calculated. First,

---

[1] CEBR grantees we interviewed cited minor issues with reporting performance data to NIJ but generally said that, over time, NIJ has clarified any reporting issues when it comes to performance measures. We also found that NIJ provides clarifying instructions to grantees through its web-based data reporting tool.

[2] CODIS has three levels where DNA profiles can be stored and searched (the Local DNA Index System – LDIS, the State DNA Index System – SDIS, and the National DNA Index System – NDIS. However, not all DNA profiles uploaded into LDIS and SDIS are searched at the NDIS level. Specifically, according to FBI officials, eligibility requirements for which DNA profiles may be uploaded and stored in LDIS, SDIS, and NDIS are based on applicable law and policy for each jurisdiction. According to officials, these laws and policies are written into CODIS software for each jurisdiction, and these rules (1) govern whether DNA profiles can be uploaded into LDIS, SDIS, or NDIS in the first place, and (2) dictate whether DNA profiles will be automatically "pushed" to the next level of CODIS (from LDIS to SDIS, and from SDIS to NDIS).

the name of the measure in the performance budget document— percent increase in the number of DNA profile uploads into the Combined DNA Index System (CODIS) from the previous fiscal year—does not clearly specify that these are DNA profiles from crime scene evidence only, rather than a combination of DNA profiles from crime scene evidence and DNA profiles from convicted offenders and arrestees. In contrast, the name of the measure in documentation provided by NIJ uses the phrase "forensic DNA profiles," which means profiles from crime scene evidence only. This is supported by NIJ documentation which shows how the measure is calculated.

Second, the name of the measure in the performance budget document says "from the previous fiscal year" but NIJ documentation shows that NIJ uses the calendar year as its basis for year-over-year comparison. Ensuring that performance measure information reported externally matches how NIJ calculates the measure would help ensure NIJ is accurately reporting results.

Measurable Target

Five of the 6 measures do not have associated targets (numerical goals). One measure—percentage increase in forensic DNA profiles uploaded to CODIS from the previous year—had a target to increase the number of forensic DNA profiles uploaded to CODIS by 7 percent in 2018. Officials said setting targets associated with the other five performance measures would not be meaningful.[3] Specifically, these measures assess outcomes associated with activities funded by CEBR grants and grantees can choose the amount of funds they spend on a variety of activities—some of which are not easily or immediately measurable, such as training lab staff and procuring equipment.[4] Thus, if NIJ were to set a program-wide target, for instance, for the "number of forensic cases analyzed," the

---

[3] NIJ officials stated that award recipients are required to provide target measures to assess their goals but NIJ does not attach targets to many of its performance measures because there are numerous uncontrollable and unpredictable factors that can contribute to erratic changes in targets on a short term and long term basis.

[4] Other activities include: salary and benefits for additional lab employees, overtime for lab staff, outsourcing, efficiency studies, and costs associated with accreditation, among others.

target might be rendered useless because grantees could choose to spend most of their money on procuring equipment that year. However, this particular challenge can be overcome by first establishing a measure that has a clear direction that indicates progress, then setting an associated target.[5] NIJ officials cited two additional reasons why they hesitate to set program-wide targets. First, officials said it is generally not NIJ's role or responsibility to set such targets. Rather, NIJ is responsible for ensuring that each CEBR grantee sets and accomplishes its individual goals and targets. Second, NIJ officials said they do not want to be held accountable for outcomes that are out of NIJ's control. For instance, NIJ cited the enactment of new laws, increases in crime rates, and technological advancements as factors outside of NIJ's control that are increasing the demand for DNA analysis. Thus, officials said, setting a program-wide target for reducing the backlog, for example, and then not meeting it may be interpreted as reflecting poorly on the CEBR program, when, in reality, there may be outside factors at play.

We agree that monitoring the performance of individual grantees is important; however, we disagree that concerns about not meeting targets due to external factors is reason to avoid setting targets. To the contrary, reviewing instances where the program missed or exceeded targets can

---

[5] We have previously reported that, prior to setting targets, managers must first create performance measures that have a clear direction that indicates progress (so that it is meaningful if the results of the measure go up or down). To do this, it may be necessary to utilize multiple data points that, understood together, indicate progress. Once a performance measure with a clear direction indicating progress has been established, managers can then set a numerical target. See GAO, *Federal Prison System: Justice Could Better Measure Progress Addressing Incarceration Challenges*, GAO-15-454, (Washington D.C.: June 2015), page 20. For example, NIJ could create a measure that assesses the "number of forensic cases analyzed" relative to the amount of funds expended for DNA analysis. Specifically, DOJ could determine a target or threshold for the number of cases analyzed while simultaneously taking into account the funding provided for DNA analysis as described in this simplified scenario: In 1 year, a grantee had a total of \$100.00 in grant funds and expended \$50.00 on DNA analysis and another \$50.00 on equipment. In that year, the grantee was able to process fifty cases, so the level of production was \$1.00 per case. During the next year, the grantee only expended \$10.00 on analysis and processed ten cases. The level of production was still \$1.00 per case. The target when taking into account funding provided could remain steady at \$1.00 per case or could aim for improved efficiencies with a target of \$.80 per case. We have not fully explored this hypothetical example and are providing it only as a means to show how two or more data points, understood together, can be used to establish a performance measure with a clear direction; and once a measure with a clear direction is established, a target can be set.

provide opportunities to identify factors that are outside of the control of the program, which if reported alongside measures, can clarify results.[6] Thus, NIJ is not accountable for results driven by external factors; rather, NIJ is responsible for explaining external factors that affect reported results. If data collected are insufficient for gauging progress, or if additional detail is needed to fully explain performance, NIJ is responsible for obtaining the necessary information to make sense of performance results.[7] Further, NIJ can set achievable targets to alleviate its concerns about being held accountable for unrealistic targets. For instance, targets related to reducing backlogs may focus on reductions in the rate of growth, which may be more achievable than focusing on reductions in size. In addition, other potential targets—such as targets related to the number or percent of grantees expected to achieve certain outputs or outcomes—may be set based on the grantees' prior trends. Therefore, setting targets related to aspects of program performance beyond CODIS uploads—for which NIJ has already established a target—would better position NIJ to demonstrate the extent to which performance is meeting expectations in other areas to Congress and other stakeholders.

Core Program Activities

As a group, the measures do not cover any core program activities. "Core program activities" means that, as a group, measures cover the

---

[6] As we have previously reported, performance reporting systems should include information to help clarify aspects of performance that are difficult to quantify or explanations for factors that were outside the control of the agency. This information is critical to understanding underlying factors that may affect reported performance. See GAO, *Information Sharing: DHS Could Better Define How it Plans to Meet Its State and Local Mission and Improve Performance Accountability*, GAO-11-223 (Washington D.C.: December 2010), p. 50. Additionally, reviewing areas where the program missed or exceeded targets can also provide an opportunity to understand factors within NIJ's control that contributed to program results. For example, in areas where the program exceeded expectations, NIJ may be able to identify and share best practices; and in areas where the program fell short of expectations, NIJ may be able to identify problems and develop corrective actions. See GAO, *Agencies' Annual Performance Plans Under the Results Act: An Assessment Guide to Facilitate Congressional Decisionmaking*, GAO/GGD/AIMD-10.1.18 (Washington, D.C.: February 1998).

[7] See GAO, *Managing For Results: Data-Driven Performance Reviews Show Promise But Agencies Should Explore How to Involve Other Relevant Agencies*, GAO-13-228, (Washington, D.C.: February 2013), p.19.

activities that an entity is expected to perform to support the intent of the program. Based on our review of "permissible uses of funds" listed in CEBR solicitations, we identified the following core program activities: adding staff or funding additional staff hours, adding equipment and supplies, providing training, outsourcing DNA analysis or testing, and other/crosscutting activities.[8] For a list of NIJ's permissible uses of funds, see appendix VIII. NIJ did not have its own list of core program activities and agreed that these uses of CEBR funds represented core program activities intended to achieve the purposes of the program. These core activities represent the strategies CEBR grantees use to achieve CEBR goals. These strategies can be used by grantees for labs that analyze crime scene DNA evidence and/or labs that test convicted offender and arrestee samples. However, NIJ's performance measures do not track the extent to which grantees use these strategies or the extent to which these tools may have an effect on desired outcomes. NIJ officials said NIJ's financial IT system—the data system that tracks grantee spending of CEBR funds— is not configured to track grantee spending by the activities listed in the "permissible uses of funds" for the purposes of performance measurement. According to NIJ, any insights that might be gained by obtaining this information—either from reconfiguring its financial IT system or through other means (such as reviewing grantee documentation after awards are closed)—would not be worth the effort to collect it. Further, NIJ reported that there are challenges to breaking down expenditures by type of DNA analysis (i.e., analysis of crime scene evidence vs. analysis of samples from convicted offenders and arrestees).

---

[8] "Other/crosscutting" activities include: contracting for DNA audits, contracting for process mapping or other efficiency studies, administrative expenses, and activities associated with and including accreditation, among other things.

However, determining a reasonable approach to measure core program activities would help NIJ have the information managers and stakeholders need to understand how the program works and how it might be improved. For example, NIJ could compare how labs spend CEBR funds to industry best practices, and thereby inform strategy for the CEBR program.[9] NIJ would not necessarily need to sort the various "permissible uses of funds" into the same groupings of "core program activities" we identified. Rather, NIJ could consider what information would be most useful for NIJ as it manages the program and for Congressional authorizers and appropriators as they oversee the program, and seek to obtain that information.

Balance

As a group, the measures cover 1 of 4 key program priorities. "Balance" means that, as a group, measures should ensure that a balance of key program priorities is covered. Based on our review of CEBR program documentation, we identified the following key program priorities: increase samples analyzed, reduce turnaround time, actual decrease in the backlog (i.e., reduction of aggregate nationwide backlog), and increase quality of analysis.[10] We found that all six performance measures relate to increasing samples analyzed.

NIJ collects data associated with 2 of the 4 program priorities—turnaround times and backlogs—from grantees, but NIJ does not consider these data to be associated with the six program-wide performance measures it uses to assess program progress, although NIJ does report these

---

[9] According to NIJ, NIJ and the Scientific Working Group on DNA Analysis Methods—a national organization that serves as a forum to discuss, share, and evaluate forensic biology methods and protocols, among other things—will be developing a publication citing lab best practices for improving efficiencies to better assist lab stakeholders across the country. According to NIJ, they aim to have this publication available by spring of 2020. To inform this effort, NIJ intends to identify CEBR award recipients that have improved their capacity, decreased their backlogs significantly, or have publicly presented their efforts to increase productivity.

[10] NIJ officials agreed with these key program priorities but clarified that our original identification of the priority of increasing the quality of analysis should be changed to "maintain or increase" the quality of analysis. They also said that the priority "actual decrease in the backlog" should be viewed in context of increasing demand for DNA analysis.

data to some extent in external reports.[11] NIJ officials stated that they do not include these additional measures as part of their official measures because there are too many confounding factors that could make turnaround time and backlogs increase. Thus, officials said, reporting such data would require too many caveats and explanations to ensure stakeholders understand that the CEBR program is achieving success.

NIJ officials also stated that one CEBR performance measure—percentage increase in forensic DNA profiles uploaded to CODIS from the previous year—reflects, among other things, CEBR contributions to enhancing the quality of DNA analysis (1 of the 4 program priorities).[12] However, we determined that, although it is possible that this measure may reflect quality improvements to some degree, it more closely relates to an increase in samples analyzed.

We recognize there are challenges to assessing the program priorities of reducing turnaround time, decreasing the actual backlog, and maintaining or increasing the quality of analysis. Without reporting on a balance of program priorities, NIJ risks overemphasizing outcomes associated with one program priority—in this case, increasing samples analyzed—over the others. Further, reporting on a balance of priorities would better position NIJ to provide information to assist Congress as it develops strategies and allocates resources to address these issues.

---

[11] See DOJ, OJP, NIJ *Report Forensic Science: Fiscal Year 2016 Funding for DNA Analysis, Capacity Enhancement, and Other Forensic Activities*, (Washington, D.C.: May 2017). See also DOJ, OJP, NIJ *Report Forensic Science: Fiscal Year 2017 Funding for DNA Analysis, Capacity Enhancement, and Other Forensic Activities*, (Washington, D.C.: April 2018).

[12] Because DNA profiles must meet a certain standard of quality to be uploaded into CODIS, NIJ reported that this measure reflects improvements in the quality of the DNA analysis. Specifically NIJ noted this measure reflects lab investments in advancing technologies and methods, training new personnel, and implementing robotics. Further NIJ reported that this measure reflects investments in quality that prevent lab shutdowns.

OJP HAS DESIGNED CONTROLS TO ACHIEVE ITS OBJECTIVES RELATED TO COMPLIANCE WITH SELECTED FEDERAL REQUIREMENTS FOR GRANTEE PROCUREMENT, BUT HAS NOT PROPERLY DESIGNED ALL CONTROLS FOR CONFLICTS OF INTEREST AND LOBBYING RELATED TO CEBR GRANTS

OJP Has Designed Controls to Achieve Its Objectives Related to Compliance with Selected Federal Requirements for Transparency in Grantee Procurement Related to CEBR Grants

According to OJP officials, grantees often use grant funds to procure lab equipment and outsource DNA analysis, among other things.[13] Procurement by grantees under CEBR awards is governed by federal regulation, which specifies that grantees that are states must use their own procurement procedures.[14] Many CEBR grantees are states; nevertheless, OJP officials said they apply controls related to transparency in grantee procurement to state grantees and to non-state grantees in the same manner.

---

[13] While there are various requirements associated with grantee procurement, we selected requirements that provide OJP or others the opportunity to review the specific details of contracts, such as the name of the contractor, how much the contract is for, and what services are being procured, among other things.

[14] 2 C.F.R. pt. 200 "Uniform Administrative Requirements, Cost Principles, and Audit Requirements for Federal Awards." Per 2 C.F.R. § 200.317, grantees that are states (including their agencies and instrumentalities thereof) must follow the same policies and procedures they use for procurements from their non-federal funds. States must comply with 2 C.F.R. § 200.322, relating to procurement of recovered materials, and ensure that every purchase order or other contract comply with any clauses required by § 200.326. A state includes any state of the U.S., D.C., the territories, and any agency or instrumentality thereof; it does not include local governments. § 200.90. All other grantees and subgrantees, including subrecipients of a state, must follow the procurement standards set forth in 2 C.F.R. §§ 318-326.

Table 3. Office of Justice Programs (OJP) Controls Related to Transparency in Grantee Procurement

Select requirements for OJP and Capacity Enhancement and Backlog Reduction (CEBR) program grantees	OJP control designed to achieve its objectives related to compliance with requirements
OJP review of procurements[a]	
As part of the grant application process, grantees must provide information about proposed procurement contracts.[b] Grantees must also provide information to the awarding agency during the award period for proposed new contracts or significant changes to existing contracts.	OJP designed a process to review all budget documentation from grantees related to proposed procurement contracts. OJP has also designed a process to review proposed new contracts or significant modifications to existing contracts during the award period.
An awarding agency may conduct pre-procurement review of grantee procurement documents when the procurement is expected to exceed the Simplified Acquisition Threshold (currently set for grantees at $250,000)[c] and is to be awarded without competition.	OJP designed has a process to review proposed sole-source contracts in excess of the Simplified Acquisition Threshold. Grantees cannot move forward with these procurement contracts until OJP provides approval of the non-competitive approach to the procurement.
Grantees are responsible for monitoring activities under federal awards to assure compliance with applicable federal requirements and performance expectations.	OJP asks grantees about monitoring mechanisms and reviews related documentation while conducting their own monitoring activities over grantees, such as during site visits.
Reporting to Congress[d]	
Beginning October 2018 and biennially thereafter, the Attorney General is to report to the Judiciary Committees the percentage of appropriated funds that each recipient—including CEBR grantees—paid to private labs to process DNA evidence, among other things.	OJP added a requirement to the fiscal year 2018 CEBR solicitation and officials said they added a special condition to the grant terms and conditions for grantees to report the amounts expended under CEBR awards on contracts for DNA analysis.

Source: GAO analysis of grantee procurement requirements and OJP information. | GAO-19-216.

[a] The first requirement comes from the CEBR grant solicitation and the DOJ Grants Financial Guide. In addition to agreeing to comply with applicable statutes and regulations, CEBR recipients agree to comply with the terms specified in the DOJ Grants Financial Guide and the solicitation. OJP outlines these requirements in the grant terms and conditions. The requirements in the guide and the solicitation are also important elements of transparency in administering CEBR grants. Other requirements come from 2 C.F.R. pt. 200, "Uniform Administrative Requirements, Cost Principles, and Audit Requirements for Federal Awards."

[b] A contract is a legal instrument by which a non-federal entity purchases property or services needed to carry out the project or program under a federal award. A contract is separate from a subaward, which is an award provided by a pass-through entity to a subrecipient for the subrecipient to carry out part of a federal award received by the pass-through entity. 2 C.F.R. §§ 200.22, 200.92.

[c] The "simplified acquisition threshold"—currently set for grantees at $250,000—means the dollar amount below which a non-federal entity may purchase property or services using small purchase methods. 2 C.F.R. § 200.88; OMB M-18-18. Small purchase procedures are those relatively simple and informal procurement methods for securing services, supplies, or other property.

[d] Requirements come from Justice for All Reauthorization Act, Pub. L. No. 114-324, § 3(b)(2)(C), 130 Stat. 1948, 1950.

We found that OJP designed one control that went beyond what was required by regulation. Specifically, as part of the grant application process, OJP designed a control to ensure that it reviews budget documentation from grantees related to all proposed procurement contracts.[15] It also has a process to review proposed new contracts, or significant modifications to existing contracts, during the award period. Table 3 provides a description of selected federal grantee procurement requirements, as well as OJP controls designed to achieve its objectives related to compliance with these requirements. For a more detailed explanation of these requirements and controls, see appendix IX.

OJP Has Designed Some Controls to Achieve Its Objectives Related to Compliance with Selected Federal Requirements for Conflicts of Interest and Lobbying in the Administration of CEBR Grants, but Has Not Properly Designed All Controls

Conflicts of Interest

Federal law prohibits government employees from participating personally and substantially in particular government matters, such as the administration of federal grants, which will affect their financial interests.[16] We identified 4 requirements related to conflicts of interest for OJP

---

[15] A contract is a legal instrument by which a non-federal entity purchases property or services needed to carry out the project or program under a federal award. A contract is separate from a subaward, which is an award provided by a pass-through entity to a subrecipient for the subrecipient to carry out part of a Federal award received by the pass-through entity. 2 C.F.R. §§ 200.22, 200.92.

[16] Criminal conflict of interest statutes governing OJP employees who administer CEBR grants are codified at 18 U.S.C. Chapter 11. Although these statutes cover a variety of topics related to conflicts of interest, our review focuses on the participation of OJP employees in government actions that may conflict with their personal financial interests, as specifically provided in 18 U.S.C. § 208. We focus on acts affecting personal financial interests because related regulatory requirements cover a broad range of issues directly applicable to OJP employees who administer CEBR grants.

employees who administer CEBR grants. We found that OJP has designed controls to achieve its objectives related to compliance with 3 of the 4 requirements and has not properly designed a control for 1 of the 4 requirements. We also found that OJP designed some controls that went beyond what was required by regulation. For example, OJP has a process for employees involved in administering CEBR grants whereby they must certify to their supervisor that they do not have conflicts of interest prior to each application cycle. In addition, in 2017 the U.S. Office of Government Ethics (OGE) cited a key DOJ resource—*the DOJ Ethics Handbook for On and Off-Duty Conduct*—as a model resource that other executive branch agency ethics programs may want to replicate.[17]

Table 4 provides a description of selected federal conflict of interest requirements applicable to OJP and its employees as they administer CEBR grants and OJP controls designed to achieve its objectives related to compliance with these requirements.[18] For a more detailed explanation of these requirements and controls, see appendix X.

As shown in the last row of Table 4, we found that OJP has not properly designed a control to achieve its objectives related to compliance with federal requirements for confidential financial disclosure. OJP designed a process to collect and review annual confidential financial disclosure reports, which are required to be filed by specified less senior employees, including those who participate personally and substantially in government actions such as grants.[19] Specifically, OJP officials stated that direct supervisors collect, review, and sign these reports. However, OJP

[17] OGE provides leadership and oversight of the executive branch ethics program, which is designed to prevent and resolve conflicts of interest. The office conducts ethics program reviews at executive branch agencies to ensure consistent and sustainable ethics program compliance with established executive branch ethics laws, regulations and policies and provides recommendations for meaningful program improvement.

[18] Because OJP administers CEBR as a formula grant program in which OJP generally allocates a specific amount of federal funds to applicants if they meet certain requirements, some risks associated with conflicts of interest that affect competitive grants, such as the appearance of or actual conflicts of interest among peer reviewers, do not apply to CEBR. Peer reviewers are subject-matter experts who provide objective, independent evaluations of competitive grant applications.

[19] 5 C.F.R. § 2634.904(a)(defining a confidential filer and providing examples of employees required to file).

does not have documentation designating which officials are authorized to certify (provide final signature) these reports, as required.

According to regulations issued by OGE, authority to certify confidential financial disclosure reports is held by an agency's Designated Agency Ethics Official (DAEO).[20]

This certification authority may be delegated to an agency's alternate DAEO or other qualified officials, such as a deputy ethics official, deputy ethics counselor, deputy standards of conduct counselor, or the equivalent. Certifiers are responsible for ensuring that reports are complete and that no interests or positions disclosed in the reports violate (or appear to violate) applicable laws and regulations.[21] OGE guidance states that agencies' written financial disclosure procedures should address which officials have been designated to review and certify reports and the levels of review and approval, among other requirements. Further, Standards for Internal Control in the Federal Government state that management should establish an organizational structure, assign responsibility, and delegate authority to achieve the entity's objectives (in this case, ensuring they are in compliance with regulations and guidance intended to help them identify the appearance of or actual conflicts of interest).[22] In doing so, management documents internal control to meet operational needs.

[20] 5 C.F.R. § 2634.605. OGE has the authority to establish a confidential financial disclosure system pursuant to section 107 of the Ethics and Governments Act, as amended, and has issued implementing regulations found at 5 C.F.R. part 2634, as well as additional guidance to executive branch agencies. The DAEO is the officer or employee who is designated by the head of an agency to administer the provisions of Title I of the Ethics in Government Act of 1978, as amended, and 5 C.F.R. part 2634 within an agency. *See* 5 C.F.R. § 2638.104 (detailing DAEO responsibilities, including review and certification of financial disclosure reports).

[21] 5 C.F.R. § 2634.605(b)(1). Effective January 1, 2019, this review will also include applicable Executive Orders.

[22] GAO-14-704G.

Table 4. Office of Justice Programs (OJP) Controls Related to Conflicts of Interest for Federal Employees

Select requirements for OJP and OJP employees	OJP control designed to achieve its objectives related to compliance with requirements
OJP employee conflict of interest requirements[a]	
Employees should not participate in matters involving specific parties, such as grants, that are likely to affect their financial interests or under circumstances that would cause a resonable person to question their impartiality.	OJP designed a process for employees involved in administering grants to review conflict of interest requirements and certify to their supervisors that they do not have conflicts of interest during each application cycle.
Employees must not engage in outside employment or other outside activities that conflict with their official duties or employment that involves certain legal practice and matters, including grants.	The Department of Justice (DOJ) has summarized applicable requirements on its ethics website the DOJ Ethics Handbook for On and Off-Duty Conduct. Employees are required to obtain written approval from the Deputy Attorney General to waive these prohibitions.
OJP ethics training requirements[b]	
New employees must complete initial ethics training and specified employees must complete annual ethics training.	OJP requires all new employees to complete initial ethics training. OJP assigns all employees to annual ethics training sessions.
OJP annual financial disclosure requirements[c]	
Senior-level employees must file public financial disclosure reports annually, and disclosures must be reviewed and certified by a designated agency ethics official. Specified less senior employees must file confidential financial disclosure reports annually, and disclosures must be reviewed and certified by a designated agency ethics official.	OJP designed a process to collect annual financial disclosure reports from all staff who administer grants. OJP's Ethics Office and OJP's Assistant Attorney General review and sign senior-level employees' reports. Less senior employees' supervisors review and sign their reports. However, OJP does not have documentation designating which officials are authorized to certify (provide final signature) these reports or the levels of review required.

Source: GAO analysis of conflicts of interest requirements and OJP information. | GAO-19-216.

[a] Requirements come from 5. C.F.R. pt. 2635, "Standards of Ethical Conduct for Employees of the Executive Branch;" 5 C.F.R. pt. 3801, "Supplemental Standards of Ethical Conduct for Employees of the Department of Justice."

[b] Requirements come from 5 C.F.R. pt. 2638, "Executive Branch Ethics Program"

[c] Requirements come from 5 U.S.C. app. §§ 101-111, "Financial Disclosure Requirements of Federal Personnel" and 5 C.F.R. pt. 2634, "Executive Branch Financial Disclosure, Qualified Trusts, and Certificates of Divestiture." Regarding the administration of CEBR grants, OJP officials stated that the Principal Deputy Assistant Attorney General for OJP, National Institute of Justice (NIJ) Director, and NIJ Principal Deputy Director positions are subject to public disclosure, and all other positions that administer CEBR grants are subject to confidential disclosure.

OJP officials said they follow guidance issued by DOJ's Justice Management Division (JMD)—led by DOJ's DAEO—which specifies the level of review that is required. Specifically, JMD's guidance states that "the component head designates employees, generally supervisors, to collect, review, and sign reports."[23] Thus, while OJP officials acknowledge they have not documented specific employee positions that are to review and certify confidential financial disclosure reports, they believe doing so would be a formality because JMD guidance only specifies that supervisory review is required.

However, per JMD guidance, the DAEO has delegated the authority to review confidential financial disclosure reports to component heads—in this case the Assistant Attorney General (AAG) for OJP—and has directed that further designation for review of such reports be made by them.

The designation of reviewers and certifiers by OJP's AAG is unclear. For instance, OJP officials said that OJP's deputy DAEO and office heads of OJP sub-components (or deputy office heads) have a role in the review process, but officials did not consistently describe those roles. Clear designations of roles within the review process is important given that OGE's guidance states that supervisors or other intermediate officials may review and sign reports, but only those who have been delegated authority may certify reports.[24]

---

[23] Procedures for the Confidential Financial Disclosure System, *available at*, https://www.justice.gov/jmd/procedures-confidential-financial-disclosure-system. *See also,* Human Resources Order DOJ 1200.1(directing component heads to designate Reviewing Officials to collect, review, and sign confidential financial disclosure reports).

[24] There are two places for signature on the reports (1) "Signature and Title of Supervisor/Other Intermediate Reviewer," and (2) Signature and Title of Agency's Final Reviewing Official" (certification). According to OGE regulation and guidance, review and signature of a supervisor or other intermediate reviewer is optional. *See* 5 C.F.R. § 2634.605 ("The reviewing official may request an intermediate review by the filer's supervisor... After obtaining any intermediate review... the reviewing official shall examine the report ...[and] certify it by signature and date.").

Table 5. Office of Justice Programs (OJP) Controls Related to Conflicts of Interest for Federal Awarding Agencies and Grantees

Select requirements for OJP employees and Capacity Enhancement and Backlog Reduction (CEBR) program grantees[a]	OJP control designed to achieve its objectives related to compliance with requirements
Awarding agencies must establish conflicts of interest policies for grants.	OJP's conflict of interest policies for grantees require grantees to agree as a condition of award acceptance to disclose conflicts of interest in writing to OJP and to maintain written standards of conduct covering conflicts of interest for grants.
Grantees must disclose in writing any conflicts of interest to the awarding agency.	OJP grant managers are to ask grantees about their policies and procedures that address conflicts of interest during grant monitoring activities, such as site visits.

Source: GAO analysis of conflicts of interest requirements and OJP information. | GAO-19-216.

[a] Requirements come from 2 C.F.R. pt. 200, "Uniform Administrative Requirements, Cost Principles, and Audit Requirements for Federal Awards."

Without documenting which employees have been delegated the authority to certify employees' confidential financial disclosure reports and the levels of review and approval required, the specific positions which have been delegated authority and the required levels of review are not clear. This may limit OJP's ability to ensure that those who review and certify reports are properly qualified to do so or that reports are reviewed by all persons OJP intends. This puts OJP at risk of failing to identify the appearance of or actual conflicts of interest among employees.

Finally, we found that OJP has designed controls to achieve its objectives related to compliance with federal requirements related to the disclosure of conflicts of interest by CEBR grantees. Table 5 provides a description of selected federal conflicts of interest requirements applicable to OJP and grantees as they administer CEBR grants and OJP controls designed to achieve its objectives related to compliance with these requirements. For a more detailed explanation of these requirements and controls, see appendix X.

Lobbying

We found that OJP has designed some controls to achieve its objectives related to compliance with federal lobbying requirements in the administration of CEBR grants but has not properly designed all controls. Federal laws contain prohibitions with respect to the use of appropriated funds for lobbying activities and violations are subject to civil penalties.[25] This includes a prohibition on recipients of federal awards using grant funds to lobby in connection with their award.

With respect to CEBR grants, we identified 4 requirements related to lobbying "certification" and "disclosure."[26] Lobbying certification refers to agreeing not to use appropriated funds to lobby. Lobbying disclosure refers to disclosing lobbying activities with respect to the covered federal action paid for with nonappropriated funds. Recipients of all federal awards over $100,000 are required to file certification documents and disclosure forms (if applicable) with the next tier above. Disclosure forms, but not certification documents, are to be forwarded from tier to tier until received by the federal agency.[27] In the case of CEBR grants, tiers include OJP,

---

[25] *See* 18 U.S.C. 1913; 31 U.S.C. § 1352. Commonly referred to as the "Anti-Lobbying Act," 18 U.S.C. § 1913 prohibits appropriated funds from being used directly or indirectly to pay for activities (e.g., advertisements, printed materials, etc.) to influence government officials to support or oppose legislation, policies, or other matters (unless expressly authorized by Congress). OJP regards this lobbying prohibition as applicable to all federal funds, including grants, although some courts have continued to apply to only to federal employees even after its scope was expanded significantly in 2002. *See, e.g., AFGE, Local 3721 v. District of Columbia,* 2005 U.S. Dist. LEXIS 8326, *33-34 (D.D.C. May 2, 2005) (reaffirming that section 1913 only applies to federal departments or agencies and their employees despite broad prohibitory language).

[26] 28 C.F.R. pt. 69, "New Restrictions on Lobbying," implements 31 U.S.C. § 1352, commonly referred to as the "Byrd Amendment." Lobbying in the context of this statute refers to paying any person for influencing or attempting to influence an officer or employee of any agency, a Member of Congress, an officer or employee of Congress, or an employee of a Member of Congress in connection with any covered Federal actions. *See also* 2 C.F.R. pt. 200, Appendix II(I), which contains similar requirements. These requirements are different from lobbying registration and disclosure requirements set forth in the Lobbying Disclosure Act of 1995, as amended (codified at 2 U.S.C. §§ 1601 *et seq.*).

[27] The required certification document is set forth in appendix A to 28 C.F.R. pt. 69. The required disclosure form is set forth in appendix B to 28 C.F.R. pt. 69. Disclosure forms are only required if the recipient has used or plans to use nonappropriated funds to lobby with respect to the award. The forwarding requirements for disclosure forms are not included on the form set forth in appendix B to 28 C.F.R. pt. 69. Additional exceptions to the prohibition and disclosure requirements apply. Pursuant to the common rule for federal agencies issued by the Office of Management and Budget and implemented by DOJ, only

grantees, subgrantees, contractors under grantees and subgrantees, and subcontractors. The specific requirements for filing lobbying certification and disclosure documents are shown in Figure 5.

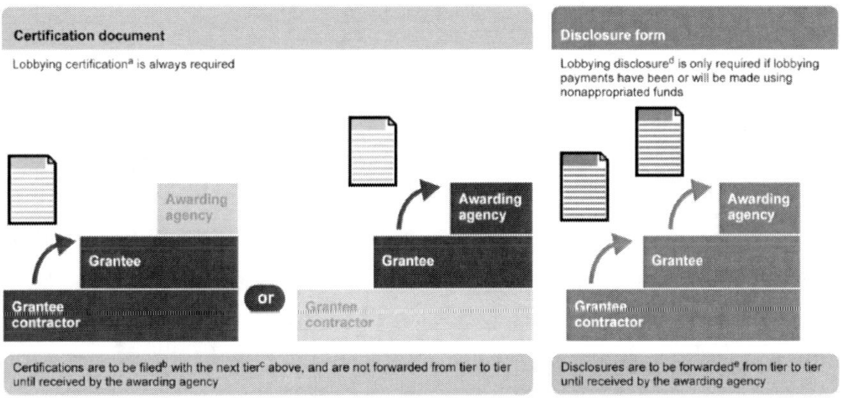

Source: GAO analysis of 28 C.F.R. pt 69 requirements for federal awards. | GAO-19-216.

Note: The Department of Justice's regulations implement the Office of Management and Budget's government-wide guidance for 31 U.S.C. § 1352, commonly referred to as the "Byrd Amendment." Lobbying in the context of this statute refers to paying any person for influencing or attempting to influence an officer or employee of any agency, a Member of Congress, an officer or employee of Congress, or an employee of a Member of Congress in connection with any covered Federal actions.

[a] Lobbying certification refers primarily to agreeing not to use appropriated funds to lobby. The required certification document also requires the same certification language in all subgrant and contract documentation.

[b] According to OJP attorneys, certifications can be "filed" by including the required certification language in award or subgrant and contract documentation. Thus, there is not a need to "file" a separate certification document.

[c] Tiers can include the awarding agency, grantees, subgrantees, contractors under grantees and subgrantees, and subcontractors.

[d] Lobbying disclosure refers to disclosing lobbying activities with respect to the covered federal action (in this case, a federal awards) paid for with nonappropriated funds. Disclosure forms are only required if the applicant, recipient, or subrecipient has used or plans to use nonappropriated funds to lobby with respect to the award. Additional exceptions to the prohibition and disclosure requirements apply.

[e] Pursuant to the regulations issued by the Department of Justice to implement the Office of Management and Budget's government-wide guidance for 31 U.S.C. § 1352, only disclosure forms, not certifications, are required to be forwarded from tier to tier until received by the awarding agency. 28 C.F.R. § 69.110(e).

Figure 5. Lobbying Certification and Disclosure Requirements for Recipients of Federal Awards over $100,000, per 28 C.F.R. pt. 69.

---

disclosure forms, not certifications, are required to be forwarded from tier to tier until received by the awarding agency. 28 C.F.R. § 69.110(e). Any person who fails to file or amend a required disclosure form is subject to a civil penalty of not less than $10,000 and not more than $100,000 for each such failure. 31 U.S.C. § 1352(c)(2).

Table 6 provides a description of select federal lobbying requirements, including the requirements in figure 5 above, as well as OJP controls designed to achieve its objectives related to compliance with these requirements. For a more detailed explanation of these requirements and controls, see appendix XI.

As shown in Table 6, we found that OJP has designed some controls to achieve its objectives related to compliance with federal lobbying requirements in the administration of CEBR grants. Specifically, we found that OJP has designed controls intended to ensure that grantees agree not to use, and agree not to allow tiers beneath them to use, appropriated funds to lobby. OJP also reviews grantees' financial reports as part of its grant approval and monitoring processes. Additionally, we found that OJP designed a control intended to ensure OJP obtains lobbying certification documents and disclosure forms from grantees. We found that OJP has not properly designed other controls to achieve its objectives related to compliance with federal lobbying requirements (see last two rows of Table 5).

Specifically, we found that, for subgrants and contracts over $100,000, OJP has not properly designed a control intended to ensure that (1) CEBR grantees obtain certification documents and disclosure forms, as applicable, from tiers below them, and (2) disclosure forms are forwarded from tier to tier until received by OJP. OJP requires grant applicants to agree to the certification document set forth in regulation.[28] This certification document, in turn, lists certification and disclosure requirements, and states that, "The Applicant shall require that the language of this certification be included in the award documents for all subgrants and procurement contracts (and their subcontracts) funded with Federal award funds and shall ensure that any certifications or lobbying disclosures required of recipients of such subgrants and procurement contracts (or their subcontractors) are made and filed in accordance with 31 U.S.C. § 1352."

[28] Appendix A to 28 C.F.R. pt. 69.

Table 6. Office of Justice Programs (OJP) Controls Related to Lobbying

Select requirements for OJP, Capacity Enhancement and Backlog Reduction (CEBR) program grantees, and others[a]	OJP control designed to achieve its objectives related to compliance with requirements
Grantees are prohibited from using appropriated funds, directly or indirectly, to lobby.[b]	As a condition of award acceptance, grantees must agree to comply with OJP's "General Conditions," which includes (1) language prohibiting the use of federal funds for lobbying, and (2) a requirement that grantees follow applicable lobbying laws, as set forth in DOJ's Grants Financial Guide. OJP also follows the grantees' use of grant funds through reviewing grantees' financial reports as part of its grant approval and monitoring processes.
Agencies are required to collect certification documents and disclosure forms[c] from grantees.[d]	OJP designed a process to collect certification documents and disclosure forms from CEBR grant applicants and grantees.
If grantees issue subgrants or contracts over $100,000, agencies are to ensure grantees collect certification documents and disclosure forms from tiers[e] below them. The agency must ensure that disclosure forms are forwarded from tier to tier until received by OJP. Grantees and subrecipients are to disclose subsequent lobbying events that require disclosure or specified events that materially affect the accuracy of previously filed disclosures at the end of each quarter.	The certification document that CEBR grant applicants agree to as part of the award acceptance process provides some information about applicable requirements. However, it does not state in clear terms what the specific requirements of the law are or how they are to be carried out, and OJP does not provide clarifying guidance.
Agencies are to take such actions as are necessary to ensure that these lobbying requirements are implemented and enforced.[f]	As stated above, OJP designed a process to collect certification documents and disclosure forms from CEBR applicants. However, it is not taking actions to ensure that grantees are requiring subrecipients to certify and disclose, and that grantees forward disclosures according to the requirements set forth above.

Source: GAO analysis of lobbying requirements and OJP information. | GAO-19-216.

[a] Unless otherwise noted, requirements are from 28 C.F.R. pt. 69, "New Restrictions on Lobbying," implementing 31 U.S.C. § 1352, commonly referred to as the "Byrd Amendment."

[b] 18 U.S.C. § 1913 prohibits appropriated funds from being used directly or indirectly to pay for activities (e.g., advertisements, printed materials, etc.) to influence government officials to support or oppose legislation, policies, or other matters (unless the activity is expressly authorized by Congress). OJP regards this lobbying prohibition as applicable to all federal funds, including grants, although some courts have continued to apply to only to federal employees even after its scope was expanded significantly in 2002. See, e.g., AFGE, Local 3721 v. District of Columbia, 2005 U.S. Dist. LEXIS 8326, *33-34 (D.D.C. May 2, 2005) (reaffirming that section 1913 only applies to federal departments or agencies and their employees despite broad prohibitory language).

[c] Disclosure forms are only required by regulation if the award recipient or subrecipient used nonappropriated funds to lobby with respect to the CEBR grant. However, we found that OJP requires all CEBR grantees to submit a disclosure form when applying for a CEBR grant.

[d] This requirement is only applicable to federal awards over $100,000. However, all CEBR grants are for at least $150,000. Additional exceptions and requirements apply.

[e] In the case of CEBR grants, tiers include OJP, grantees, subgrantees, contractors under grantees and subgrantees, and subcontractors.

[f] 31 U.S.C § 1352; see also 28 C.F.R. § 69.410.

However, OJP's certification document does not state in clear terms what the specific requirements of the law are or how they are to be carried out, and OJP does not provide guidance to grantees to clarify the requirements. Additionally, we found that OJP does not follow up with grantees to ensure they are implementing the requirements to obtain lobbying certification documents and disclosure forms, as applicable, from tiers beneath them, and forwarding disclosure forms to OJP.

Standards for Internal Control in the Federal Government state that management should (1) externally communicate the necessary quality information to achieve the entity's objectives (in this case, compliance with this regulation), and (2) implement control activities through policies (in this case, following up with grantees to ensure they are implementing the requirements).[1]

Although referenced in OJP and DOJ documentation, OJP officials responsible for overseeing the implementation of these lobbying requirements were not fully aware of specific legal requirements. In particular, they were not aware that tiers beneath grantees were required to file certification documents and disclosure forms, as applicable, with tiers above them; nor were they aware that disclosure forms were required to be forwarded from tier to tier until received by OJP for subgrants and contracts over $100,000.[2] According to OJP officials, this was due, in part, to OJP focusing primarily on ensuring that grant funds are not used to lobby.[3] Officials said that ensuring grant funds are not used to lobby is more important than ensuring non-grant funds used to lobby are disclosed. Moreover, OJP officials said, the specific requirements in the regulation pertaining to certification and disclosure of lobbying activities at tiers below grantees are confusing and take significant effort to understand.[4]

---

[1] GAO-14-704G.

[2] As discussed above, OJP and DOJ have published information about the relevant lobbying regulation in agency documents, such as the lobbying certification form that recipients must execute prior to receiving award funds and the *DOJ Grants Financial Guide* which grantees agree to comply with as a condition of receiving award funds.

[3] As noted in the Table 5, OJP has designed controls to (1) ensure grantees agree not to use grants funds to lobby, and (2) review grantees' financial reports as part of its grant approval and monitoring processes.

[4] 28 C.F.R. pt. 69.

In 2009, the DOJ Office of the Inspector General recommended NIJ "establish procedures to ensure that the required lobbying disclosure forms are submitted for all grantees, subgrantees, and contractors, and that the disclosures are considered when evaluating grant applications for award."[5] However, we found that OJP did not properly design controls intended to ensure that grantees obtain lobbying disclosure forms from subgrantees and contractors, as appropriate, and forward them to OJP, as required.

At least one other agency has taken steps to clarify federal lobbying certification and disclosure requirements. In 2012, the Department of Transportation's Federal Transit Administration identified that its award recipients may not have been correctly applying these requirements. To address this issue, the Federal Transit Administration (1) issued guidance for its award administrators to remind award recipients of the requirements, and (2) requested that administrators ensure award recipients review their activities and take steps to ensure full compliance.[6]

Clearly communicating requirements to grantees would better position OJP to ensure that grantees and tiers beneath them (subgrantees and contractors) are fully aware of these requirements. For instance, while not generalizable, 3 out of the 4 CEBR grantees we spoke with were not aware of one or more of these requirements. As a result, OJP, grantees, and tiers beneath them may not be fully complying with all federal lobbying disclosure requirements. Further, CEBR subgrantees and contractors could be using their own non-grant funds to engage in lobbying activities related to CEBR grants that neither the grantee nor OJP are aware of.

---

[5] Department of Justice, Office of the Inspector General, *U.S. Department of Justice Audit of the National Institute of Justice's Practices for Awarding Grants and Contracts in Fiscal Years 2005 through 2007,* Audit Report 09-38 (Washington, D.C.: September 2009). The text we use to quote this recommendation comes from the Executive Summary of the report. Notably, this same recommendation, as stated on page 41 of the report, does not reference contractors. We used the text from the recommendation in the Executive Summary because OJP restated it, and agreed with it, in its formal response to the audit (see appendix IX of the report). The OIG closed the recommendation as implemented in 2011 as OJP designed controls intended to ensure OJP obtains lobbying disclosure forms from grantees.

[6] Federal Transit Administration, *Certifications and Disclosure of Lobbying Activities* (Washington, D.C.: August 17, 2012), www.transit.dot.gov/funding/granteeresources/certifications-and-assurances/certifications-and-disclosure-lobbying, accessed October 15, 2018.

Conclusion

While DOJ has awarded nearly $1 billion to CEBR and its legacy programs since 2004, as of 2017, about 169,000 requests for crime scene DNA analysis were backlogged at state and local government crime labs. Backlogs persist for various reasons, including scientific advancements that enable law enforcement to obtain investigative leads from smaller amounts of biological evidence.

NIJ has not consistently documented CEBR program-wide goals and CEBR performance measures do not fully reflect attributes that would help NIJ assess progress toward those goals. Thus, NIJ is limited in its ability to communicate intended program results and progress toward those results. This information could inform Congress as it seeks to address challenges in the area of DNA evidence and allocate resources to address DNA evidence backlogs, capacity enhancement, and other priorities.

Further, while questions have been raised about potential improper connections among those who profit from DNA analysis and those who advocate for CEBR funding, we found that OJP has designed controls to achieve its objectives related to compliance with selected federal requirements associated with transparency in grantee procurement. However, OJP has not documented which employees have been delegated authority to certify employees' confidential financial disclosure reports or clarified all applicable federal requirements associated with lobbying. Addressing these issues would increase transparency into how OJP identifies and addresses the appearance of or actual conflicts of interest.

Recommendations for Executive Action

We are making the following four recommendations to OJP:

The Principal Deputy Assistant Attorney General for OJP should consistently document CEBR program-wide goals to clarify intended program results. (Recommendation 1)

The Principal Deputy Assistant Attorney General for OJP should ensure that performance measures for each CEBR program-wide goal fully reflect appropriate attributes of successful performance measures. (Recommendation 2)

The Principal Deputy Assistant Attorney General for OJP should document which employee positions have been delegated certification (final signature) authority for confidential financial disclosure reports and specify required levels of review and approval. (Recommendation 3)

The Principal Deputy Assistant Attorney General for OJP should (1) clarify its guidance to grantees to specify what their requirements are under 28 C.F.R. pt. 69 with regard to obtaining lobbying certification documents, and obtaining and forwarding to OJP lobbying disclosure forms, from tiers beneath them; and (2) design a control to follow-up with grantees to help ensure they are meeting these requirements. (Recommendation 4)

AGENCY COMMENTS AND OUR EVALUATION

We provided a draft of this chapter to DOJ for review and comment. In written comments, which are reproduced in full in appendix XII, DOJ concurred with the four recommendations and described actions planned to address them. DOJ also provided technical comments, which we have incorporated as appropriate.

DOJ concurred with our first recommendation to consistently document CEBR program-wide goals. Specifically, DOJ's OJP stated that it will review how these goals are communicated to grant recipients, Congress and stakeholders to determine what changes may be needed to ensure that the goals are consistently communicated.

DOJ also agreed to address our second recommendation on improving performance measures. OJP stated that it plans to develop a logic model that will more clearly indicate for grant recipients the measures and quantitative goals for their projects that may be achieved with CEBR funds. This model could be useful to grantees and could help address our recommendation if it establishes clear linkages between program activities, program-wide performance measures, and program-wide goals.

In addition, DOJ agreed to address our third recommendation related to documenting which employee positions have been delegated certification (final signature) authority for confidential disclosure reports. OJP stated that it would adopt its own version of a policy implemented by DOJ's Justice Management Division, which specifies that supervisors can sign reports. We believe this action, if implemented, would address our recommendation if OJP's version of the policy included a clear designation of which employee positions have been delegated certification (final signature) authority and specified required levels of review and approval.

Finally, DOJ concurred with our fourth recommendation to clarify guidance to grantees regarding their requirements under 28 C.F.R. pt. 69, and to design a control to ensure grantees are meeting these requirements. Specifically, OJP stated that in February 2019, it implemented a centralized standard lobbying certification procedure so all award applicants will provide their certification at the point of registration. In addition, OJP stated that, beginning with the fiscal year 2019 grant awards, OJP plans to update a checklist it uses as part of its grant monitoring. According to OJP, this update will help OJP ensure that all applicable lobbying disclosure forms—including those from tiers beneath grantees—are collected and submitted to OJP. We look forward to reviewing these actions in more detail to determine if they meet the intent of our recommendation.

We are sending copies of this chapter to the appropriate congressional committees, the Attorney General, and other interested parties.

Gretta L. Goodwin
Director, Homeland Security and Justice

APPENDIX I: OBJECTIVES, SCOPE, AND METHODOLOGY

This chapter addresses the following questions:

1) What is known about the amount of backlogged crime scene DNA evidence, including sexual assault kits (SAK), in state and local government labs and the factors that contribute to such backlogs?
2) What is known about the amount of unsubmitted DNA evidence, including SAKs, in law enforcement custody and the factors that contribute to this unsubmitted evidence?
3) To what extent does the Office of Justice Programs (OJP)—the primary grant-making arm of DOJ—measure DNA Capacity Enhancement and Backlog Reduction grant program (CEBR) performance?
4) To what extent has OJP designed controls related to conflicts of interest, transparency in grantee procurement, and lobbying requirements applicable to CEBR grants?

DNA Evidence Backlog Data (Objective 1)

To describe what is known about the amount of backlogged crime scene DNA evidence in state and local labs, we reviewed data from (1) the Bureau of Justice Statistic's (BJS) Census of Publicly Funded Forensic Crime Labs,[7] (2) West Virginia University's Project FORESIGHT,[8] and (3) DOJ's CEBR grant program. We determined that the most useful data for showing aggregate nationwide trends was CEBR data.[9] Regarding CEBR

---

[7] DOJ's BJS, a component within OJP, surveys all publicly-funded crime labs in the U.S. periodically in order to provide aggregate information on services provided and resources required (according to BJS, the last survey covered 2014). This includes information on the number of requests for crime scene DNA analysis and convicted offender and arrestee samples received, the number of requests and samples completed, and the number of requests and samples backlogged.

[8] Project FORESIGHT is a research project designed to help crime labs evaluate their efficiency and inform resource allocation decisions. It is run out of West Virginia University's College of Business and Economics and uses data from participating national, state, and local labs. Data submitted from labs include data on DNA analysis workloads and backlogs, among other things. According to a representative of Project FORESIGHT, as of October 2018, 149 lab systems in the U.S. (143 government lab systems representing 275 facilities and 6 private lab systems representing 8 facilities) contributed data to the project.

[9] BJS's Census of Publicly Funded Forensic Crime Labs occurs periodically (e.g., 2005, 2009 and 2014) and is thus not useful for showing year-over-year trends. Participation in Project FORESIGHT has increased over time; however, determining the extent to which it is

data, we collected yearly "baseline" data from the National Institute of Justice (NIJ), the component within DOJ that administers the CEBR program, for calendar years 2011 through 2017.[10] We selected these years because in 2011 two previous grant programs were combined into what is now known as the CEBR grant program, and because 2017 is the most recent full calendar year for which CEBR grantee data were available at the time of our review.

To assess the extent to which NIJ-compiled CEBR baseline data we use in this chapter are reliable, we completed a number of data reliability steps, including discussing data entry issues and data limitations with NIJ and select grantees; running logic tests on the data, verifying which grantees reported data each year (2011-2017), and comparing NIJ-compiled baseline data to a limited amount of source data from grantee reports, and comparing CEBR data against other data (BJS survey data for 2014 only). After completing data reliability steps, we determined that the CEBR baseline data we use in this chapter are sufficiently reliable for our purpose, which is to show year-over-year trends in workloads and backlogs among state and local government labs.

We determined that state and local government labs participating in the CEBR program from 2011 through 2017 generally represent the level of workloads and backlogs from state and local government labs that participate in CODIS nationwide.[11] However, we found that CEBR

---

generalizable to government labs in the U.S. for each year for which we desired to show trend data would have taken significant resources. In contrast, grantee participation in CEBR has been relatively consistent over time, and almost all state and local government labs are represented in CEBR data.

[10] NIJ collects baseline data—which includes data from all labs within grantees jurisdictions (not only those labs that use CEBR grant funds)—as part of the grant application process.

[11] As of January 1, 2018 there were 194 state and local government labs in the U.S. that performed DNA analysis on evidence from crime scenes and participated in the FBI's Combined DNA Index System (CODIS). According to information provided by OJP, also as of January 1, 2018, CEBR grantees represented 189 of these 194 state and local government labs. For the 5 labs not included, we determined that they constituted a small portion of backlog data in other years so as not to substantively affect results (for example, less than 3 percent since 2012). We were unable to identify the number of labs that did not participate in the CEBR program for years 2011 through 2016. However, NIJ officials said that CEBR grantees have generally been representative of all state and local government labs in the U.S. that participated in CODIS in prior years as well. In addition, to ensure that fluctuations in the population of CEBR grantees during these years did not significantly

turnaround time data for 2017 were not reported as consistently as data for workloads and backlogs and may not be representative of nationwide turnaround times for state and local government labs. Thus, in the report, we provide an example of the types of turnaround times reported by CEBR grantees, which may provide insights into turnaround times at state and local government labs across the United States.

To ensure we had identified all possible sources of data and to better understand how labs collect and report data, we met with a non-generalizable selection of knowledgeable officials and representatives from 11 entities, including 2 national associations, 2 academic research groups, 5 CEBR grantees (or grantee labs), and 2 components within DOJ.[12] We selected the national association because it represents labs across the country; we selected CEBR grantees that differed by type of jurisdiction (state versus local) with one CEBR grantee also being a grantee of DOJ's Sexual Assault Kit Initiative grant program; we selected the academic groups based on literature we had reviewed and a recommendation from DOJ officials; and we selected NIJ because it administers the CEBR program and BJS because it has data on lab outcomes.

affect our results, we compared data on the number of requests backlogged and the annual rate of change in the number of requests backlogged for each year from all grantees (including those grantees who reported data from some but not all years between 2011 and 2016) to data from a subset of grantees (including those that provided data for all years between 2011 through 2016). We found that any fluctuations did not affect overall conclusions about trends in DNA analysis in these prior years.

[12] Specifically, we met with representatives from the American Society of Crime Lab Directors and the International Association of Chiefs of Police. We also met with selected CEBR grantees or grantee labs, including: Washington D.C. Department of Forensic Sciences; Dallas County, TX, Southwestern Institute of Forensic Science; Harris County, TX, Institute of Forensic Sciences; Houston, TX, Houston Forensic Science Center; and Maryland State Police. We also met with academic researchers from West Virginia University's Project FORESIGHT as well as authors of a 2010 report entitled *Unanalyzed Evidence in Law Enforcement Agencies: A National Examination of Forensic Processing in Police Departments*. Finally, we met with federal officials from OJP's NIJ and BJS.

Unsubmitted DNA Evidence, Including SAKs, in Law Enforcement Custody (Objective 2)

To identify and describe what is known about the amount of unsubmitted DNA evidence, including SAKs, in law enforcement custody, we reviewed information and/or data related to several efforts to inventory or quantify this evidence, including SAKs, in law enforcement custody.

- DOJ's National Sexual Assault Kit Initiative (SAKI) grant program. We obtained SAKI data from OJP's Bureau of Justice Assistance (BJA). BJA provided us with data from grantees that covered the time period of October 2015 through June 2018. We obtained data for this timeframe because the SAKI grant program began in October 2015, and June 2018 was the latest period for which SAKI data were available at the time of our analysis. To ensure the reliability of these data, we performed electronic checks for missing or duplicate data entries among grantees, and compared BJA-manipulated (i.e., "cleaned") data to raw SAKI data to ensure that the only discrepancies between the two datasets were those discrepancies identified by BJA officials. We also reviewed related documentation, such as survey questions and instructions for reporting data, and interviewed BJA officials and representatives from a SAKI contractor. We found these data to be sufficiently reliable for our purposes, which are to describe the extent to which SAKI (1) has contributed to an understanding of the number of unsubmitted SAKs in law enforcement possession among participating jurisdictions, and (2) has reported on activities and outcomes associated with analyzing previously unsubmitted SAKs.
- New York County District Attorney's Office (DANY) SAK grant program. We obtained data from DANY's Sexual Assault Kit Backlog Elimination Grant Program from DANY officials. DANY officials provided us with aggregated grantee data that covered the time period of October 2015 through September 2018. We

obtained data for this timeframe because it reflects the first and last quarters from which data were available at the time of our analysis. To ensure the reliability of these data, we obtained and reviewed the forms and spreadsheets DANY used to obtain these data from grantees and aggregate them. We also interviewed DANY grant program administrators to understand the processes they use to aggregate grantee data. We found these data to be sufficiently reliable for our purpose, which is to describe the extent to which DANY grants have contributed to an understanding of the number of unsubmitted SAKs in law enforcement possession among participating jurisdictions.

- State SAK inventory data. We obtained publicly-available aggregated SAK inventory data from Idaho and Texas websites. To select these states, we first conducted legal research to identify states with laws requiring them to conduct inventories of SAKs where evidence had not yet been submitted for lab analysis. We selected Texas and Idaho as illustrative examples because they differed with respect to legal requirements for conducting inventories, and timeframes associated with conducting those inventories.[13] To ensure the reliability of these data, we interviewed state officials about their data collection efforts and steps they took to ensure reliability; we also followed up with officials to ensure the data we present are accurate and up to date. We found these data to be sufficiently reliable for our purpose, which is to provide illustrative examples of how state laws contribute to understanding the number of unsubmitted SAKs in law enforcement possession.
- Reporting requirements under the Sexual Assault Forensic Evidence Reporting (SAFER) Act. We reviewed the reporting requirements for grantees and DOJ under the SAFER Act of

[13] Texas law required a one-time inventory by October 2011 and Idaho law required a onetime inventory by December 2016, to be followed by annual inventories.

2013.[14] However, no data had been reported under this Act at the time of our review.

- Joyful Heart Foundation. We reviewed the website where the Joyful Heart Foundation posts links to the sources of data it uses to quantify the number of unsubmitted SAKs in law enforcement custody in state and local jurisdictions nationwide. We also interviewed Joyful Heart Foundation representatives responsible for collecting and aggregating these data to discuss their methods for ensuring the data are reliable. However, given the many and varied sources of data, we did not independently perform steps to ensure each source was sufficiently reliable.
- Academic research. We reviewed two prior nationwide studies on the amount and reasons for unsubmitted DNA evidence in law enforcement custody (including DNA evidence associated with various types of crime, not just sexual assault). These two studies were the only studies we were able to identify that assessed the amount of unsubmitted DNA evidence on a nationwide scale. However, we did not include findings from either of these studies in our report because the data supporting these studies were not current.[15]

Lastly, we identified challenges associated with inventorying SAKS within jurisdictions and quantifying SAKs across jurisdictions based on (1)

---

[14] The SAFER Act of 2013, Pub. L. No. 113-4, § 1002, 127 Stat. 54, 127-131, added a new purpose area and related requirements to the Debbie Smith DNA Backlog Grant Program (34 U.S.C. § 40701). This Act authorizes grants for the purpose of conducting audits of sexual assault evidence, and requires the Attorney General to publish information from these audits online.

[15] One study provided nationwide estimates of homicide, rape, and property crime cases with forensic evidence that was not submitted to a lab. See Strom, Kevin J. et. al., *The 2007 Survey of Law Enforcement Forensic Evidence Processing* (Washington, D.C.: 2009). See https://www.ncjrs.gov/pdffiles1/nij/grants/228415.pdf, accessed Oct. 23, 2018. Another study provided nationwide estimates of homicide, rape, and property crime cases with possible biological evidence which local law enforcement agencies had not submitted to a lab. See Lovrich, Nicholas P. et.al., *National Forensic DNA Study* Report, Final Report (Washington, D.C.: 2004), https://www.ncjrs.gov/pdffiles1/nij/grants/203970.pdf, accessed Oct. 23, 2018.

qualitative survey responses that DANY officials provided us;[16] (2) relevant DOJ reports; and (3) interviews with DNA evidence stakeholders.

Factors that Contribute to DNA Evidence Backlogs and Unsubmitted DNA Evidence, Including SAKs (Objectives 1 and 2)

To identify and describe factors that contribute to backlogs of unanalyzed DNA evidence at labs and unsubmitted DNA evidence, including SAKs, in law enforcement custody, we reviewed 22 reports, including 16 government reports (or government-funded reports), 4 academic journal articles, 1 book, and 1 study from a non-governmental organization. Five of the government (or government-funded) reports included nationwide studies of public crime labs and law enforcement agencies with original research. We identified this collection of literature by conducting database searches of peer reviewed material, government reports, and conference papers using the Online Computer Library Center and ProQuest Professional database search engines. We used keywords including "DNA evidence," "backlog," "law enforcement," "laboratories," and "awaiting testing," among others. We discussed these factors with DNA evidence stakeholders from 17 entities, including: 2 national associations; 6 grantees of the CEBR, SAKI, and or DOJ's Sexual Assault Forensic Evidence—Inventory, Tracking, and Reporting (SAFE-ITR) grant programs; 2 states' law enforcement agencies; 4 academics and practitioners; 2 components within DOJ, and 1 additional federal agency.[1711] We selected these entities based on their knowledge of crime

---

[16] As part of the regular reporting DANY required of its grantees, DANY asked that grantees respond to the following question: "Have the kits that will be tested under this grant been fully inventoried?" For those grantees that responded "no," the questionnaire asked "If no, how is your inventory process progressing?" DANY officials provided us with grantees responses to this question over time without identifying individual grantees by name. Grantees' responses to this question provided insights into challenges they faced in performing SAK inventories.

[17] Specifically, we met with representatives from the following national associations: the American Society of Crime Lab Directors and the International Association of Chiefs of Police. We also met with select state and local labs that are grantees of the CEBR, SAFE-

scene DNA evidence collection, storage, and analyses; as well as their familiarity with DOJ grant programs that address crime scene DNA evidence. Finally, we conducted legal research on state laws that require law enforcement to submit SAKs for testing and that also require labs to analyze previously unanalyzed SAKs. We summarized information from these sources to identify common factors, and we included illustrative examples of the types of factors we identified in this chapter.

CEBR Grant Program Performance Measurement (Objective 3)

To evaluate how DOJ measures CEBR program performance, we first sought to identify CEBR program-wide goals. To do this, we reviewed OJP CEBR documentation, including the most recent CEBR grant solicitation and NIJ reports that include CEBR performance information. We also discussed CEBR goals with NIJ officials. We then assessed the CEBR program-wide goals against federal internal control standards that call for management to define goals clearly.[18]

To evaluate the extent to which CEBR program-wide performance measures reflect attributes of successful performance measures, we obtained and reviewed OJP documentation that defined the six CEBR program-wide performance measures NIJ currently uses to assess CEBR program-wide performance. This documentation included definitions of measures, the methodology used to calculate the measures, and any

---

ITR, and/or SAKI grant programs, including: Dallas County, TX, Southwestern Institute of Forensic Sciences; Washington D.C. Department of Forensic Sciences; Georgia Criminal Justice Coordinating Council; Harris County, TX, Institute of Forensic Sciences; Houston Forensic Science Center; and Maryland State Police. We also met with representatives from the Idaho State Police and Texas Department of Public Safety. We also met with academics and practitioners, including the director of West Virginia University's Project FORESIGHT, authors of a 2010 report entitled *Unanalyzed Evidence in Law Enforcement Agencies: A National Examination of Forensic Processing in Police Departments,* the District Attorney of New York County, and a retired law enforcement officer. Additionally, we spoke to representatives from a survivors' advocacy group (the Joyful Heart Foundation). Finally, we met with federal officials from DOJ's NIJ and BJA, as well as officials from the Congressional Research Service.

[18] GAO, *Standards for Internal Control in the Federal Government*, GAO-14-704G (Washington, D.C.: September 2014), Principle 6 – Define Objectives and Risk Tolerances (starts on page 35).

associated targets. To determine the extent to which DOJ's six CEBR program-wide performance measures effectively assessed progress, we compared them against 5 of 9 attributes of successful performance measures identified in our previous work.[19] We selected 3 of the 9 attributes (linkage, clarity, and measurable targets) because they are foundational. By "foundational" we mean that, without them, other attributes are less relevant or important. We selected 2 of the 9 attributes (core program activities, balance) because they assess the extent to which the performance measures cover a variety of aspects of performance.[20] To assess the extent to which NIJ's six CEBR performance measures met these criteria, two GAO analysts independently assessed each performance measure against these 5 criteria, and then met to discuss and reconcile differences. A GAO subject matter expert on performance measures provided input on the design, execution, and reporting of this audit objective.

---

[19] Our prior work establishes 9 attributes of successful performance measures: linkage, clarity, measurable targets, objectivity, reliability, core program activities, limited overlap, balance, and government-wide priorities. GAO, *Tax Administration: IRS Needs to Further Refine Its Tax Filing Season Performance Measures*, GAO-03-143 (Washington, D.C.: November 2002).

[20] We excluded 4 of the 9 attributes. Specifically, we excluded 3 of the 9 attributes (objectivity, reliability, and limited overlap) because they may not be relevant if performance measures are not aligned (linked) with program goals; they may be difficult to understand without clarity across measures; and they may lack meaning if the measures do not have targets and thus do not set performance expectations. We excluded 1 of the 9 attributes (government-wide priorities) because we determined that it is not necessarily applicable to the CEBR program. The attribute of "government-wide priorities" specifies that performance measures should cover a range of priorities, such as quality, timeliness, efficiency, cost of service, and customer satisfaction. While important, we believe it is not realistic to expect that a single grant program include performance measures that cover most or all government-wide priorities. Rather, we believe that attributes of "balance" and "core program activities"—which we did include—are sufficient to ensure that the performance measures cover an appropriate variety of aspects of performance.

OJP Controls Related to Conflicts of Interest, Transparency in Grantee Contracting, and Lobbying Applicable to CEBR Grants (Objective 4)

To evaluate the extent to OJP has designed controls related to conflicts of interest, transparency in grantee procurement, and lobbying requirements applicable to CEBR grants, we conducted legal research in these topic areas to identify federal statutes and regulations applicable to OJP and CEBR applicants, recipients, and subrecipients.[21] We also reviewed DOJ documentation such as the DOJ Ethics Handbook for On and Off-Duty Conduct, OJP Grants Management Manual, and the DOJ Grants Financial Guide to identify any additional requirements (including requirements that DOJ imposes on itself which may not stem from statute or regulation). OJP attorneys verified that the list of requirements we had compiled was complete and accurate. To determine the extent to which OJP designed controls to achieve its objectives related to compliance with these requirements, we reviewed the Fiscal Year 2018 CEBR solicitation, DOJ Departmental Ethics Website, DOJ Ethics Handbook for On and Off-Duty Conduct, DOJ Grants Financial Guide, OJP ethics training materials, and OJP's "General Conditions" for grants (available on OJP's website). To further determine the extent to which OJP has designed each control, we also obtained and reviewed a variety of CEBR program documents, including those associated with reviewing and approving CEBR grant applications, and grant monitoring. Examples of these documents include: CEBR grantee budget narratives and budget detail worksheets (submitted as part of a grant application), OJP's checklist for ensuring grantee application packages have all required information, and OJP's grant monitoring checklists (which OJP grant managers use when conducting grant monitoring activities).[22] We also discussed some of the controls that

---

[21] We examined requirements applicable to subgrantees and contractors under CEBR grantees and subgrantees, and subcontractors.

[22] Grant monitoring allows grant managers to observe compliance with requirements and progress against project goals, and ensure that adequate controls are in place to improve accountability. Grant monitoring activities are carried out through communication with the

we identified with four of the CEBR grantees discussed earlier. We did not review the extent to which controls were implemented effectively (e.g., through controls testing procedures such as conducting case file reviews), nor did we review the extent to which employees were trained to implement controls effectively (e.g., through reviewing training material and/or outcomes of such training). Finally, we assessed the controls we identified as not properly designed against federal internal control standards which specify how management should design controls to achieve its compliance objectives.[23]

We conducted this performance audit from September 2017 to March 2019 in accordance with generally accepted government auditing standards. Those standards require that we plan and perform the audit to obtain sufficient, appropriate evidence to provide a reasonable basis for our findings and conclusions based on our audit objectives. We believe that the evidence obtained provides a reasonable basis for our findings and conclusions based on our audit objectives.

APPENDIX II: ADDITIONAL SOURCES OF LAB DATA AND CONSIDERATIONS WHEN REVIEWING DNA EVIDENCE DATA

In addition to reviewing the DNA Capacity Enhancement and Backlog Reduction (CEBR) grant program data, we also reviewed data from (1) the Bureau of Justice Statistic's (BJS) Census of Publicly Funded Forensic Crime Labs, and (2) West Virginia University's Project FORESIGHT. We determined that the most useful data for showing aggregate nationwide trends was CEBR data and used CEBR data in our report.[24] In reviewing

grantee, desk reviews (reviews of documents in grantees' files), and in-depth monitoring (consisting of site visits and enhanced programmatic desk reviews).

[23] GAO-14-704G. Per guidance in these standards, an existing control is not properly designed when, even if the control operates as designed, the control objective would not be met.

[24] We also presented CEBR data in a testimony statement in July 2018. See GAO, *DNA Evidence: Preliminary Observations on DOJ's DNA Capacity Enhancement and Backlog Reduction Grant Program*, GAO-18-651T (Washington D.C.: July 18, 2018).

these datasets and through discussions with lab officials, we identified a number of considerations related to analyzing and interpreting data from any of these sources, or from individual labs. These considerations may pose difficulties assessing data across labs or datasets. However, we did not find that these considerations invalidated the CEBR data we present in the report.

Additional Sources of State and Local Lab Data

BJS Census of Publicly Funded Forensic Crime Laboratories

BJS, a component within the Office of Justice Programs (OJP), surveys all publicly-funded crime labs in the U.S. periodically in order to provide aggregate information on services provided and resources required (according to BJS, the last survey covered 2014). This includes information on the number of requests for crime scene DNA analysis and convicted offender and arrestee samples received, the number of requests and samples completed, and the number of requests and samples backlogged.

Project FORESIGHT

Project FORESIGHT is a research project designed to help crime labs evaluate their efficiency and inform resource allocation decisions. It is operated through West Virginia University's College of Business and Economics and uses data from participating national, state, and local labs. Data submitted from labs include data on DNA analysis workloads and backlogs, among other things. According to a representative of Project FORESIGHT, as of October 2018, 149 lab systems in the U.S. (143 government lab systems representing 275 facilities and 6 private lab systems representing 8 facilities) contributed data to the project. A recent software project, called FORESIGHT 20/20, enables participating labs to provide data, and receive reports back from project FORESIGHT, almost automatically (with a few mouse clicks).

Considerations When Reviewing Lab DNA Evidence Data

Identifying the Universe of Labs

Data on DNA-related workloads and backlogs may be reported at different levels. Specifically, data may be reported by a single lab facility, a multi-lab system (containing more than one lab facility), or a grantee (which also may represent more than one lab facility). Additionally, data sets may include data from government labs only, or government labs and private labs. Further, DNA analysis of crime scene evidence and DNA testing of convicted offender and arrestee samples involve different processes, and some labs only perform one function or the other.

Using "Cases" versus "Requests"

Labs or datasets may refer to "requests" for DNA analysis, or they may refer to these requests for DNA analysis as "cases." We use "requests" in this chapter, since there may be several requests associated with a law enforcement "case."

Counting Requests for DNA Analysis as One Request or Two Requests

Requests for DNA analysis may be defined and tabulated differently. In our report, we define DNA analysis as (1) biology screening (locating, screening, identifying, and characterizing blood and other biological stains and substances); and/or (2) DNA testing (identifying and comparing DNA profiles in biological samples). However, one data collection effort we reviewed and one lab we interviewed considered requests for biology screening and requests for DNA testing as separate requests. Specifically, we found that reported data on DNA requests may count requests in two ways:

1) Biology screening and/or DNA testing as one request. If the request requires biology screening and/or DNA testing, it is counted once as one request (such that if a request requires one or both, it is counted once)

2) Biology screening and DNA testing as separate requests. If the request requires only biology screening, it is reported as a "request for biology screening;" if the request requires only DNA testing, it is counted as a "request for DNA testing;" and if the request requires both biology screening and DNA testing, it is counted as two requests—one request for biology screening and one request for DNA testing.

Including or Excluding Requests Closed by Administrative Means

Data sets may or may not include requests closed by administrative means in reported data. Some requests may be closed by administrative means rather than through analysis, such as when a suspect pleads guilty before the evidence is analyzed or when the victim has not consented to participate in the criminal justice process. These requests may or may not be included in the number of requests received or the number of requests completed. A related issue occurs when law enforcement officials or prosecutors fail to withdraw requests from the lab's queue when DNA analysis is no longer needed.

Including or Excluding Outsourced Requests

Data sets may or may not include requests that government labs outsourced to other labs for DNA analysis (including private labs). A related issue occurs when local law enforcement agencies outsource DNA analysis directly to private labs. In these instances, the DNA evidence is shipped from the law enforcement agency directly to the private lab, which performs DNA analysis. After the private lab performs DNA analysis, it provides the results of the analysis to a government lab for review and potential upload of the results into the FBI's Combined DNA Index System (CODIS).[25]

[25] CODIS is a system that allows federal, state, and local labs to exchange and compare DNA profiles electronically in order to develop investigative leads. Only federal, state, or local government labs that meet the FBI's Quality Assurance Standards can participate in CODIS.

Calculating and Interpreting the Meaning of Turnaround Time for Requests

Labs or data sets may calculate turnaround times for requests differently, and interpreting turnaround time may require additional explanatory information to help understand how labs are performing, as described in the bullets below.

- Including/excluding "expedited" requests: Data on turnaround time may or may not include expedited requests, which occur when law enforcement officials or prosecutors mark a request as needing to be expedited or rushed due to various factors.
- Different start and end points: Labs or data sets may start or end the "clock" for measuring turnaround time at different points. For instance, according to two lab directors we interviewed, some labs may begin the clock for counting turnaround time when the area of the lab that performs DNA analysis takes custody of the evidence, and other labs may not begin the clock until an analyst begins working on the request.[26] NIJ defines the ending of the clock for turnaround time as when the lab issues a report with the results to law enforcement. Lab directors also suggested there could be different ending points for turnaround time.
- Interpreting the link between turnaround time and performance: According to NIJ officials, a high average turnaround time may not necessarily be an indicator of poor performance at a lab or across lab populations. This is because turnaround time is a measurement that looks backward at completed requests for analysis. Thus, a lab that has made a concerted effort to clear its backlog of old requests for analysis will, in the immediate aftermath, report a higher

[26] Evidence associated with a given request may need to first be processed in one or more other areas of the lab—such as areas that perform latent fingerprint analysis or firearms analysis—before being processed by the area of the lab that performs DNA analysis. Thus, another potential source of confusion is that turnaround time for a given request can be thought of as (1) the time it takes to process evidence only in the area of the lab that performs DNA analysis, or (2) the time it takes to process evidence in the area of the lab that performs DNA analysis and in other areas of the lab where evidence may also need to be processed.

average turnaround time. Labs may also complete requests associated with violent crimes faster than requests associated with nonviolent crimes, which may or may not be reported separately.

Analyzing "Requests" versus More Granular Information

Depending on the purpose of the analysis, it may be helpful to review data that goes beyond "requests." According to NIJ and stakeholders we spoke with, all requests are not equal and do not take the same amount of resources. Specifically, requests for analysis may contain one or more "items" for examination, such as weapons, carpets, or bedsheets. Each item may contain multiple "samples" for analysis (e.g., multiple stains on clothing). Each sample, in turn, may be subject to multiple "tests," such as instrumental analysis, extractions, or comparative examinations. Thus, depending on the number of items, samples, and tests needed, requests can vary greatly in the amount of resources needed.

Appendix III: Information on Backlogs of Convicted Offender and Arrestee Samples

The processing of convicted offender and arrestee samples involves the DNA testing of the samples and the subsequent review and upload of the resulting DNA profiles into the FBI's Combined DNA Index System (CODIS). Data from the National DNA Index System (NDIS)—the national level of CODIS—show that convicted offender and arrestee sample DNA profile uploads into CODIS increased significantly from 2003 through 2010, then have generally decreased since 2010, as shown in Figure 6.[27]

[27] CODIS has multiple levels where DNA profiles can be stored and searched: the local level, the state level, and the national level. NDIS is the national level of CODIS.

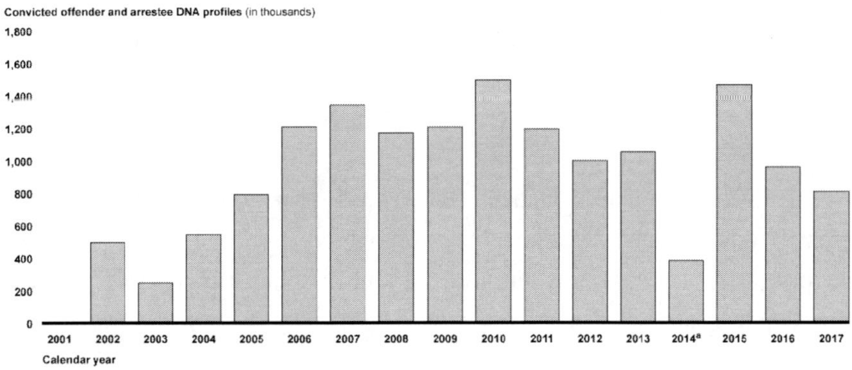

Source: GAO analysis of Federal Bureau of Investigation data. | GAO-18-216.

Note: A DNA profile is generally considered to be a unique genetic identifier based on the genetic constitution of an individual. DNA samples may be taken from convicted offenders, arrestees, and other categories of persons as authorized by law.

[a]The data show a significant drop in profiles uploaded in 2014, followed by a large increase in 2015. FBI officials said this was the effect of a 2014 legal decision in California that forced state officials to remove many offender profiles from CODIS. Shortly after, California was able to reinstate the profiles, which explains the large increase in 2015.

Figure 6. Convicted Offender and Arrestee DNA Profile Uploads into the National DNA Index System (NDIS).

According to NIJ, a primary reason for the growth in convicted offender and arrestee DNA profile uploads prior to 2010 was the passage of state laws requiring the collection and testing of DNA samples. For instance, NIJ reported that by 2009, the federal government and all 50 states had passed bills requiring collection of DNA from offenders convicted of certain crimes; in addition, the federal government and many states had also passed legislation to allow collection from people who are arrested for certain crimes. According to NIJ, between 2005 and 2010, it made more than $58 million available to reduce the backlog of samples of convicted offenders and arrestees. According to officials from NIJ, the FBI, and the American Society of Crime Lab Directors, this funding helped build labs' capacities such that convicted offender and arrestee backlogs are no longer an issue of concern.

Appendix IV: Grant Programs That Address DNA Evidence Backlogs and Unsubmitted Sexual Assault Kits

The Department of Justice (DOJ) has two grant programs that primarily address DNA evidence backlogs at labs and two grant programs that primarily address unsubmitted sexual assault kits (SAK) in law enforcement custody.

Grants That Primarily Address DNA Backlogs at Labs

The DNA Capacity Enhancement and Backlog Reduction grant program (CEBR) dates back to 2004 and funds are available to address DNA evidence backlogs in labs, among other things. The CEBR program is funded by an appropriation "for a DNA analysis and capacity enhancement program and for other local, State, and Federal forensic activities."[28] The broad appropriations language enables the National Institute of Justice (NIJ) to allocate this funding for a variety of forensic programs. However, the Justice for All Reauthorization Act of 2016 mandated that at least 75 percent of funds made available under this appropriation be used for grants for DNA analysis or to increase the capacity of government labs to carry out DNA analysis.[29]

---

[28] The appropriation language states that funds are "for a DNA analysis and capacity enhancement program and for other local, State, and Federal forensic activities, including the purposes authorized under section 2 of the DNA Analysis Backlog Elimination Act of 2000 (Public Law 106–546) (the Debbie Smith DNA Backlog Grant Program)." The purposes of the CEBR grant program are generally similar to the purposes of the Debbie Smith DNA Backlog Grant Program. There is no additional statutory authorization for the program.

[29] The Justice for All Reauthorization Act of 2016, Pub. L. No. 114-324, § 3(a), requires that not less than 75 percent of the funds made available under this appropriation be provided for grants for activities described under paragraphs (1), (2), and (3) of section 2(a) of the DNA Analysis Backlog Elimination Act of 2000. Those purposes include (1) To carry out, for inclusion in the Combined DNA Index System of the FBI, DNA analyses of samples collected under applicable legal authority; (2) To carry out, for inclusion in such Combined DNA Index System, DNA analyses of samples from crime scenes, including samples from rape kits, samples from other sexual assault evidence, and samples taken in cases without an identified suspect; and (3) To increase the capacity of laboratories owned by the State or by

Table 7. Department of Justice (DOJ) Grant Programs that Address DNA Evidence Backlogs at Labs

Grant program	Year began	Fiscal year 2018 participation and amount awarded[a]	Description
DNA Capacity Enhancement and Backlog Reduction (CEBR) Program[b]	2004	127 awards $67.8 million awarded	CEBR is a formula[c] grant program administered by the National Institute of Justice (NIJ). Eligible participants are states and units of local government with existing crime labs. Award funds support capacity building and analysis at labs related to (1) DNA evidence collected from crime scenes, and (2) DNA samples taken from convicted offenders and arrestees.
Forensic DNA Lab Efficiency Improvement and Capacity Enhancement (EI&CE) Program	2017	19 awards $9.3 million	EI&CE is a grant program administered by NIJ. Awards are made competitively. Eligible participants are states and units of local government with existing crime labs. Applications must be project based. Award funds may be used to support efficiency and capacity-building at labs related to DNA and non-DNA disciplines (such as forensic analysis of latent fingerprints and firearms).

Source: GAO analysis of DOJ documents. | GAO-19-216.

[a]These amounts reflect amounts awarded in fiscal year 2018, with the actual project period beginning January 1, 2019.

[b]The CEBR program has several legacy names dating back to 2004. In 2011, grant programs that separately funded labs that (1) analyzed crime scene DNA evidence and (2) tested convicted offender and arrestee DNA samples, were combined into one grant program called the DNA Backlog Reduction Program. In 2014, this was renamed the DNA Capacity Enhancement and Backlog Reduction Program.

[c]Formula grant awards are made non-competitively to states and units of local government based on a formula set by DOJ (as opposed to a formula set by statue in the case of some grant programs) that allocates certain amounts to each state. This formula takes into account each state's population and associated crime levels, and guarantees a minimum amount for eligible applicants from each state.

---

units of local government to carry out DNA analysis of samples specified in the purposes above. Prior to this, congressional reports accompanying the appropriation have directed that the Office of Justice Programs (OJP) make funding for DNA analysis and capacity enhancement a priority. OJP officials said they use CEBR funding to help meet the "75 percent" requirement discussed above.

In 2017, NIJ began the Forensic DNA Laboratory Efficiency Improvement and Capacity Enhancement (EI&CE) grant program which is intended to complement the CEBR program by allowing more flexibility in how funds can be used.[30] Both CEBR and EI&CE program details are in Table 7.

Grants That Primarily Address Unsubmitted SAKs in Law Enforcement Custody

The Sexual Assault Kit Initiative (SAKI) program began in 2015 and addresses unsubmitted SAKs and other related challenges. SAKI is administered by the Bureau of Justice Assistance (BJA), a component within OJP which provides leadership and services in grant administration.[31]

In 2016, NIJ began the Sexual Assault Forensic Evidence – Inventory, Tracking, and Reporting (SAFE-ITR) program, which provides funding to inventory SAKs, track their movement, and report on their status (e.g., in law enforcement custody, submitted to labs, etc.).[32] These grant programs are detailed in Table 8.

---

[30] The EI&CE program is funded by the same appropriation as the CEBR program, and OJP officials said they also use funding provided through this grant program to meet the "75 percent" requirement discussed in the previous footnote. The CEBR program does not permit the use of funds for non-DNA disciplines. According to DOJ, enhancing capacity and improving efficiency in the processing and testing of non-DNA evidence from cases that also involve a request for DNA analysis will ultimately reduce the backlog of DNA evidence. According to NIJ officials, this is because evidence associated with a given request may need to first be processed in one or more other areas of the lab—such as areas that perform latent fingerprint analysis or firearms analysis—before being processed by the area of the lab that performs DNA analysis. Thus, according to officials, enhancing capacity and improving efficiency in other areas of the lab will result in a shorter overall lab processing time for the request as a whole.

[31] According to OJP officials, SAKI is authorized and funded by an appropriation "for a grant program for community-based sexual assault response reform." *See* Consolidated Appropriations Act, 2017, Pub. L. No. 115-31, 131 Stat. 135, 204.

[32] According to OJP the SAFE-ITR program for fiscal years 2016 and 2017 was authorized and funded by appropriations "for a grant program for community-based sexual assault response reform," and/or "a DNA analysis and capacity enhancement program and...forensic activities." *See id.* Program requirements for fiscal years 2016 and 2017 for the SAFE-ITR program are described in NIJ's grant solicitations and are similar, but not identical, to the new authorized grant program area contained in the SAFER Act of 2013, Pub. L. No. 113-4,

Table 8. Department of Justice (DOJ) Grant Programs that Address Sexual Assault Kits (SAK) in Law Enforcement Custody that Have Not Been Submitted to Labs for DNA Analysis

Grant program	Year began	Fiscal year 2018 participation and amount awarded[a]	Description
Sexual Assault Kit Initiative (SAKI)	2015	32 awards $42.9 million	• SAKI is a grant program administered by DOJ's Bureau of Justice Assistance. Awards are made competitively. • Eligible participants include state law enforcement agencies and units of local government, among other entities. • Award funds support multidisciplinary community response teams that inventory, track, and analyze previously unsubmitted SAKs, and perform a variety of other related services.[b] Award funds may also be used to collect lawfully-owed DNA samples and support investigation and prosecution of cold case sexual assaults.[c]
Sexual Assault Forensic Evidence – Inventory, Tracking, and Reporting (SAFE-ITR)	2016	3 awards $1.4 million	SAFE-ITR is a grant program administered by DOJ's National Institute of Justice. Awards are made competitively. Eligible participants are states and units of local government. • Award funds support efforts to inventory, track, and report on previously unsubmitted SAKs as they move from collection through final disposition.

Source: GAO analysis of DOJ documents. | GAO-19-216.

[a] These amounts reflect amounts awarded in fiscal year 2018, with the actual project period beginning on October 1, 2018 for SAKI grantees and January 1, 2019 for SAFE-ITR grantees.

[b] These other services may include, for example, producing necessary protocols and policies in support of improved coordination and collaboration among labs, police, prosecutors, and victim service providers; or providing resources to address sexual assault investigations and prosecutions.

[c] Lawfully-owed DNA samples are DNA samples that have never been collected from convicted offenders, arrestees, and potentially others, even though applicable law allows for their collection. According to BJA, collection of lawfully-owed DNA samples and investigation and prosecution of cold case sexual assaults should be undertaken only after a jurisdiction has made significant progress in addressing issues associated with unsubmitted SAKs.

§ 1002, 127 Stat. 54, 127 (amending 34 U.S.C. § 40701). For fiscal year 2018, DOJ is required by the Justice for All Reauthorization Act of 2016 to allocate no less than 5 percent of funds authorized and funded by appropriations for "a DNA analysis and capacity enhancement program and...forensic activities" to this new purpose areas, in addition to other requirements Pub. L. No. 114-324, § 3, 130 Stat. 1948, 1949. According to OJP officials, the fiscal year 2018 SAFE-ITR grant program fulfills this allocation requirement.

APPENDIX V: SELECTED EFFORTS TO INVENTORY OR QUANTIFY UNSUBMITTED SEXUAL ASSAULT KITS AND ASSOCIATED CHALLENGES

The information listed below provides descriptions of, and data from, large-scale efforts to inventory or quantify unsubmitted sexual assault kits (SAK) in law enforcement custody. For some efforts, the data provided goes beyond counts of unsubmitted SAKs in law enforcement custody, and provides downstream outputs and outcomes associated with the previously unsubmitted SAKs—including counts of SAKs analyzed, and associated CODIS hits, criminal investigations, and prosecutions.

There are challenges to obtaining reliable data on unsubmitted SAKs, discussed below. Nevertheless, we took steps to ensure the data we present below are sufficiently reliable for our purpose, which is to provide data that accurately reflect the activities and outcomes of the large-scale efforts we discuss.[33] We also note that data from these efforts may overlap, which means that it is not possible to combine totals from each effort without potentially duplicating results.

National Sexual Assault Kit Initiative (SAKI) Grantee Data

SAKI grants provide funding to state law enforcement agencies and units of local government to help them address unsubmitted SAKs. The SAKI program requires grantees to first take an inventory of unsubmitted SAKs; however, SAKI grantees may also use SAKI funds to test SAKs and pursue investigations, among other things.[34]

---

[33] See our scope and methodology in appendix I for the specific steps we took.

[34] According to BJA documentation, SAKI aims to create a coordinated community response to achieve just resolution to sexual assault cases by (1) establishing or strengthening a comprehensive and victim-centered approach, (2) capacity building to prevent high numbers of unsubmitted SAKs in the future, and (3) supporting the investigation and prosecution of cases for which SAKs were previously unsubmitted. More information on the SAKI program is in appendix IV.

Table 9. Counts of Unsubmitted Sexual Assault Kits (SAK) Inventoried, Analyzed, and Outcomes from Analysis and Investigation as Reported by 38 SAKI Grantees (October 2015 – June 2018)

	Prior to SAKI[a]	SAKI-funded[b]	Total
Collection and Storage			
Number of SAKs Inventoried (a SAKI requirement)	70,611	53,929	124,540
Number of unsubmitted SAKs Identified	59,614	43,223	102,837
Submission and Prioritization			
Number of previously-analyzed SAKs[c]	10,997	10,705	21,702
Number of SAKs determined not to require DNA analysis[d]	2,225	5,544	7,769
Number of SAKs determined to require DNA analysis	49,847	30,991	80,838
Number of SAKs submitted for DNA analysis	26,196	42,484	68,680
Lab Analysis and Reporting			
Number of SAKs tested to completion	-	33,228	33,228
Number of uploads to the Federal Bureau of Investigation's Combined DNA Index System (CODIS)[e]	9,599	11,336	20,935
Number of CODIS hits[f]	6,110	5,001	11,111
Number of CODIS hits to known serial sex offenders	-	832	832
Investigation and Prosecution			
Potential Investigations	-	4,602	4,602
Potential Prosecutions	-	3,682	3,682
Charges Resulting in Plea Bargains	-	378	378
Post-Trial Convictions	-	93	93

Source: GAO analysis of National Sexual Assault Kit Initiative (SAKI) grantee data. | GAO-19-216.

Note: Blank entries indicate data from prior to the SAKI program that SAKI did not ask for or were not able to obtain from grantees.

[a] These data reflect outputs and outcomes from activities that took place by jurisdictions before the first allocation of SAKI funding. According to BJA, the SAKI program does not collect data for some pre-SAKI activities in the "Lab Analysis and Reporting" and "Investigation and Prosecution" categories, which are reflected by blank cells. As such, total activities in these instances should be considered representative of SAKI-funded activities only. BJA officials stated that they believe the actual totals for these categories would be larger if such data were available.

[b] SAKI-funded data includes only those outputs and outcomes from activities that took place as a result of SAKI funds. The SAKI program does not allow grantees to use SAKI funds to inventory SAKs that were collected after the grant application date.

[c] Previously-analyzed SAKs may appear in an inventory because the kit was discovered to have been analyzed only after it was included in the inventory. However, a previously-tested SAK may be determined to require testing under SAKI if it was only partially tested in the past—such as only subject to biological screening or analyzed using non-CODIS eligible DNA methodologies.

[d] A SAK may be determined not to require analysis if, for example, the victim chose not to participate in the criminal justice process or law enforcement determined through the facts of the case that a crime did not occur.

[e] CODIS is a database that allows federal, state, and local labs to exchange and compare DNA profiles electronically. As a result of processing evidence from crime scenes, only DNA profiles believed to belong to a potential perpetrator can be uploaded into CODIS.

[f] The FBI defines a hit as "a confirmed [DNA profile] match that aids an investigation and one or more of the case(s) involved in the match are unsolved."

BJA collects SAKI performance data from grantees related to these SAKI-funded activities, as well as data on activities grantees completed prior to receiving SAKI funds. BJA officials who administer the SAKI program stated that they believe the number of CODIS hits, investigations, prosecutions, and convictions connected to many of these previously unsubmitted SAKs will continue to increase over time as these outputs and outcomes often take time to materialize.

New York County District Attorney's Office (DANY) Grantee Data

In 2015, DANY awarded nearly $38 million in grants to 32 jurisdictions (grantees) across 20 states to help them analyze SAKs.[35] DANY and SAKI program administrators coordinated to ensure that the DANY and SAKI programs complemented each other and did not duplicate resources. Funding awarded through DANY's SAK Program is only allowed to be used for costs associated with analyzing SAKs; thus, DANY did not provide funds for grantees to perform inventories.[36] Nevertheless, DANY tracks the number of SAKs submitted to labs for analysis, which DANY officials said serves as a proxy for previously unsubmitted SAKs. In addition, DANY collects data on outputs and outcomes that occur as a result of SAK analysis. According to DANY officials, the number of investigations, prosecutions, and convictions will increase following the end of the grant program as these outcomes often

---

[35] According to DANY documentation, New York City was the first major jurisdiction to make a comprehensive effort to eliminate previously unsubmitted SAKs. According to DANY documentation, between 2000 and 2003, New York City sent out approximately 17,000 SAKs for DNA analysis. DANY officials we spoke with said that when New York County— one of the 5 counties in New York City—received windfall money from a legal settlement, the county wanted to use the money to help other jurisdictions replicate what New York City had done.

[36] The DANY grant solicitation did not distinguish between "unsubmitted" SAKs in law enforcement jurisdiction and "backlogged" SAKs in laboratories. Specifically, according to the DANY SAK Program grant solicitation, a SAK eligible for analysis with grant funds is one that is connected to a reported sexual assault that has not been analyzed within 365 days of being booked into law enforcement evidence regardless of the reason it had not been analyzed.

take time to materialize. Though the DANY grant program ended in September 2018, grantees will continue to share results for another year.

Table 10. Counts of Sexual Assault Kits (SAK) Submitted for Analysis, and Outcomes from Analysis and Investigation, as Reported by 32 Grantees Participating in the New York County District Attorney's Office (DANY) SAK Grant Program (as of September 2018)

	DANY program data
Collection and storage	
This step is outside of the scope of the DANY program	
Submission and prioritization	
Number of SAKs submitted for DNA analysis[a]	62,915
Lab analysis and reporting	
Number of uploads to the Federal Bureau of Investigation's Combined DNA Index System (CODIS)[b]	16,657
Number of CODIS hits[c]	8,185[d]
Investigation and prosecution	
Arrests	196
Prosecutions	184
Convictions	58

Source: GAO presentation of DANY's Sexual Assault Kit Backlog Elimination Program data. | GAO-19-216.

Note: Fields under headings "Lab analysis and reporting" and "Investigation and prosecution" represent grantee-reported outputs and outcomes resulting from DANY-funded SAK analysis. DANY funds were not used to directly assist grantees with these activities.

[a] Data collected by DANY grant administrators do not include the number of unsubmitted SAKs identified. These data do, however, provide a count of the number of SAKs submitted to labs for analysis, which DANY officials said serves as a proxy for previously unsubmitted SAKs.

[b] CODIS is the generic term used to describe the FBI's program of support for criminal justice DNA databases as well as the software used to run these databases. It enables federal, state, and local labs to exchange and compare DNA profiles electronically, thereby linking crimes to each other and to known convicted offenders and arrestees.

[c] The FBI defines a hit as "a confirmed [DNA profile] match that aids an investigation and one or more of the case(s) involved in the match are unsolved."

[d] Of these 8,185 CODIS hits, DANY labeled 2,785 of them as "confirmatory" hits (more of the 8,185 hits may be confirmatory hits since, at the time of our review, two of the grantees had not reported to DANY how many of their total hits were confirmatory hits). According to DANY, confirmatory hits reflect instances where the suspected perpetrator was identified and potentially convicted of the associated sexual assault prior to SAK analysis. Thus, SAK analysis and the related CODIS profile match confirm the known suspect or perpetrator. Some of these confirmatory hits may meet the FBI's definition of a CODIS hit, and others may not (see the FBI's definition of a hit in table note "c"). For instance, according to FBI officials, CODIS matches that confirm the identity of known suspects in unsolved cases—where such confirmation through DNA analysis had not already occurred—qualify as hits; however, CODIS matches that identify known perpetrators in solved cases do not qualify as hits.

Data from State Efforts to Address Unsubmitted SAKs

Table 11. Data Resulting from Laws Requiring Inventories of Unsubmitted Sexual Assault Kits (SAK) in Idaho and Texas

State	Summary of law	Inventory data
Idaho	One-time inventory by December 2016 and annual inventory beginning January 2017	• December 2016:1,116 unsubmitted SAKs inventoried, 541 of which required analysis[a] • Update through end of 2017: 463 additional SAKs were collected during 2017 (making a total of1,572 SAKs from both 2016 and 2017); of these, 509 were submitted to a lab and 316 were analyzed by a lab during 2017 • Update through September 2018: 354 additional SAKs were collected from January 2017 through September 2018 (making a total of 1,933 SAKs since the one-time inventory in 2016); of these, 541 were submitted to a lab during this time and 218 were analyzed during this time
Texas	One-time inventory by October 2011[b]	• February 2013:15,823 sexual assault cases with unsubmitted evidence inventoried[c] • Update through August 2017: 3,132 additional sexual assault cases with unsubmitted evidence inventoried, making a total of 18,955 cases; of these, 16,773 have been analyzed and 2,142 remain to be analyzed.

Source: Analysis of state laws, reports, and interviews with state officials. | GAO-19-216.

[a] Under the Idaho code, a SAK is not required to be analyzed if (1) there is no evidence to support a crime being committed, (2) it is no longer being investigated as a crime, or (3) an adult victim expressly indicates that no further forensic examination or testing occur. Idaho Code. Ann. § 67- 2919(8).

[b] This law became effective on September 1, 2011. 2011 Tex. Gen. Laws 1105, § 18. Guidance provided by the Texas Department of Public Safety (TDPS) to law enforcement agencies instructed them to provide TDPS with data on evidence collected prior to July 31, 2011. A report was due from TDPS by February 2013.

[c] The law required law enforcement agencies to inventory "all active criminal cases for which sexual assault evidence had not yet been submitted for laboratory analysis," as opposed to an inventory of SAKs. 2011 Tex. Gen. Laws 1105, § 15(a)(1). According to a senior official from TDPS, law enforcement primarily reported cases that included SAKs. Additionally, TDPS provided guidelines to law enforcement agencies that operationalized the statutory definition of "active criminal cases." See Tex. Gov't Code Ann. § 420.003. Specifically, the guidance states that (1) the offense has to have been reported to law enforcement and the evidence must be in the custody of the law enforcement agency or an accredited crime lab, and (2) this does not apply to sexual assault evidence that was collected prior to September 1, 1996.

As of September 2018, we identified at least 26 states with laws that require the conducting of SAK inventories where evidence had not yet been submitted for lab analysis.[37] Twenty-four of these laws were passed in

[37] Legislation varies by state. For example, some states require one-time inventories and other states require annual inventories; additionally, some states require inventories of SAKs and others require inventories of sexual assault cases (which often include SAKs). Other variations may apply.

2014 or later. We provide data on unsubmitted SAKs identified in Idaho and Texas in Table 11 as illustrative examples of the results of these types of laws.

Reporting Requirements under the Sexual Assault Forensic Evidence Reporting (SAFER) Act

The SAFER Act of 2013 authorizes grants for the purpose of conducting audits of sexual assault evidence, and requires the Attorney General to publish information from these audits online.[38] According to OJP officials, because no appropriations were required to be allocated for grants under this new grant purpose area until 2018, OJP was under no obligation to collect or report on this information until that time.[39] However, according to OJP officials, now that an appropriation has been made for grants for this new purpose area, OJP has taken steps to modify an existing grant program to ensure it complies with related requirements.[40] Specifically, according to OJP officials, NIJ made changes to the Sexual Assault Forensic Evidence – Inventory, Tracking, and Reporting (SAFE-ITR) grant program to ensure it fully complies with the additional statutory requirements of the SAFER Act.[41]

---

[38] The SAFER Act of 2013, Pub. L. No. 113-4, § 1002, 127 Stat. 54, 127-131, added a new purpose area and related requirements to the Debbie Smith DNA Backlog Grant Program (34 U.S.C. § 40701).

[39] Specifically, according to OJP officials, until 2018, NIJ grantees were under no statutory obligation to submit the reports described at 34 U.S.C. § 40701(n)(4) that would have been required consistent with the SAFER Act of 2013 amendments to the Debbie Smith DNA Backlog Grant Program statutory authorization.

[40] DOJ is required by the Justice for All Reauthorization Act of 2016 to allocate no less than 5 percent of funds authorized and funded by appropriations for "a DNA analysis and capacity enhancement program and...forensic activities" for the new purpose area authorized by the SAFER Act of 2013. Pub. L. No. 114-324, § 3, 130 Stat. 1948, 1949. The Justice for All Reauthorization Act also contained additional requirements related these activities; in particular, the funds must also be used to create and operate associated tracking systems and to prioritize cases in which the statute of limitations will soon expire.

[41] The SAFE-ITR grant program began in 2016. According to OJP officials, the SAFE-ITR program for fiscal years 2016 and 2017 was authorized and funded by appropriations "for a grant program for community-based sexual assault response reform," and "a DNA analysis and capacity enhancement program and...forensic activities." According to these officials, program requirements for fiscal years 2016 and 2017 for the SAFE-ITR program are

The intent of this program is to support states and local governments as they inventory, track, and report on previously unsubmitted SAKs.[42]

OJP officials said they plan to begin publishing information online from fiscal year 2018 SAFE-ITR grantees, but they said the information will be limited at first because grantees will have just begun their award periods (begins January 1, 2019).

The SAFER Act requires that grantees provide reports to DOJ every 60 days, and that DOJ publish these reports online within 7 additional days.[43] Per the Act, these reports are to include, for each grantee, among other things, the cumulative total number of samples of sexual assault evidence that, at the end of the reporting period that: (1) are in the possession of the state or unit of local government, (2) have been submitted to a lab, (3) have been analyzed to completion.[44] OJP officials said that potential SAFE-ITR grantees may consider these and other reporting requirements (not listed here) to be burdensome, especially when compared to the reporting requirements of SAKI grants.[45] As evidence of this sentiment among potential SAFE-ITR grantees, OJP officials stated that, in fiscal year 2018, they only awarded $1.4 million out of a total of $5.5 million available to SAFE-ITR grantees, due to a lack of interest among applicants. In contrast, in fiscal year 2018, the SAKI program received more applications for federal funding than they had available (they awarded $42.9 million).

The SAFER Act also allows for optional reporting of sexual assault forensic evidence information from non-grantee states and local governments. Specifically, the Act requires that the Attorney General make the reporting form that SAFE-ITR grantees use available to all states and units of local government. States and local governments can then, at their sole discretion, submit reports to DOJ for publication.[46]

---

described in NIJ's grant solicitations and are similar, but not identical, to requirements added by the SAFER Act of 2013 to the Debbie Smith DNA Backlog Grant Program statutory authorization.

[42] We include more information on SAFE-ITR in appendix IV.

[43] 34 U.S.C. § 40701(n)(4)(A),(C).

[44] 34 U.S.C. § 40701(n)(4)(B).

[45] SAKI grants also allow funds to be used to conduct inventories of unsubmitted SAKs, and also require grantees to report data associated with those inventories.

[46] 34 U.S.C. § 40701(n)(4)(E)(ii).

OJP reported that it plans to have the SAFE-ITR reporting form finalized by January 1, 2019, the start date for fiscal year 2018 SAFE-ITR grantees.

Joyful Heart Foundation Project to Aggregate State and Local Efforts to Inventory or Quantify Unsubmitted SAKs

The Joyful Heart Foundation, a survivors' advocacy organization, has ongoing work to quantify the number of unsubmitted SAKs in law enforcement custody nationwide. According to representatives from Joyful Heart, as of October 2018, they had obtained data from 35 states and 32 cities or counties.[47] The Joyful Heart Foundation reported that it obtained data made available through legislative mandates, executive actions, or audits required under federal funding requirements. The Foundation reported that it also obtained data through its own public records requests. They publish the data and provide links to their sources on a publicly-available website. Representatives said they do their best to ensure the sources they obtain data from are reliable, the data is not double-counted, and the data are up to date. Nevertheless, given the many and varied sources of government and other publicly-available data used to calculate the figure, we did not independently perform steps to ensure each source was sufficiently reliable for our purposes.

Challenges to Inventorying SAKs within Jurisdictions and Quantifying SAKs across Jurisdictions

We identified several challenges associated with inventorying SAKs within jurisdictions and quantifying them across jurisdictions.[48]

---

[47] Joyful Heart Foundation representatives said that in instances where they obtain data on unsubmitted SAKs in cities or counties that are located in states where they have obtained statewide data, they adjust the state data to avoid double-counting.

[48] For information about the steps we took to identify these challenges, see our scope and methodology in appendix I.

These challenges make it difficult to obtain complete and accurate counts at the jurisdiction level, and contribute to the difficulty of obtaining a definitive count of unsubmitted SAKs nationwide. The challenges we identified include the following:

- Difficulty locating SAKs. DOJ has reported that inconsistencies in how and where SAKs are stored may pose difficulties for SAK tracking and management. DOJ also reported that SAKs may be in law enforcement property rooms; crime labs; or in rape crisis centers, hospitals, or other medical facilities where they were originally collected. Additionally, DNA evidence stakeholders we interviewed said that a significant challenge to inventorying SAKs is the act of physically locating them.
- Insufficient or inconsistent information technology (IT) systems. NIJ has reported that the quality of law enforcement evidence tracking systems may be a significant challenge in many law enforcement agencies across the country.[49] Additionally, BJA officials stated that the multitude of disparate and sometimes incompatible data systems used within and across jurisdictions can make tracking and identifying a SAK difficult.[50]
- Lack of incentive or awareness. According to NIJ-funded research on unsubmitted SAKs, law enforcement may be hesitant to provide counts of these SAKs since reporting the number of unsubmitted SAKs publicly may be seen as a reflection of poor performance or may expose law enforcement to legal action.

---

[49] For example, when NIJ began a project to help the city of Detroit, Michigan, address its SAK backlog, NIJ reported that Detroit's police department property room "database" was a simple spreadsheet, and it was not easy to figure out which SAKs might have been analyzed over the years.

[50] For example, there may be different data systems used by the lab, the police department, and the district attorney's office within a single jurisdiction and these differences are magnified when working across different jurisdictions.

- Representatives from the Joyful Heart Foundation suggested that this may also reflect a lack of priority given to sexual assault cases. Additionally, BJA officials stated that some SAKI grantees have found that certain law enforcement jurisdictions may not be aware that they had unsubmitted SAKs in their custody.
- Resource constraints. Some stakeholders we interviewed stated, and NIJ has reported, that when law enforcement agencies perform inventories, they do not always have the needed funding, personnel, or training and technical expertise to provide complete and accurate counts. However, OJP officials and survivor advocacy representatives said that, generally speaking, jurisdictions were able to mobilize resources to perform inventories of unsubmitted SAKs once they made it a priority to do so.
- Understanding which SAKs should be included in inventories. There is no single definition of what qualifies as an 'unsubmitted SAK.' For instance, depending on the effort, an inventory of SAKs may or may not include SAKs that are past the statute of limitations or that were collected before a certain time period. An inventory also may or may not include unreported SAKs.[51]
- Ensuring counts are reliable and up to date. The number of unsubmitted SAKs is constantly changing as (1) law enforcement agencies locate or identify old SAKs in storage, (2) SAKs are submitted to labs and are therefore no longer considered "unsubmitted," and (3) new SAKs are booked into evidence. Thus, counts or inventories are a snapshot of unsubmitted SAKs at a given time and require constant monitoring and updating.

[51] According to NIJ, an unreported SAK is a SAK from a victim who has consented to the collection of the SAK but has not consented to participate in the criminal justice process. Also according to NIJ, an unreported SAK cannot be submitted to a laboratory for analysis, unless applicable law allows.

Appendix VI: DNA Capacity Enhancement and Backlog Reduction Program-Wide Performance Measures

Table 12. DNA Capacity Enhancement and Backlog Reduction (CEBR) Program-wide Performance Measures, Detailed Summary

Performance measure	Definition (according to the National Institute of Justice)	Years in use since 2011	Data source	Scope of performance measure
Crime scene DNA analysis				
Number of Forensic Cases Analyzed[a]	Performance measure captures the number of cases that grantees analyzed using DNA CEBR grants, including all outsourced cases.	2011-2017	CEBR grantee performance reports (aggregated)	Lab activities funded by CEBR grants
Number of Forensic DNA Profiles[b] Uploaded to the FBI's Combined DNA Index System (CODIS)[c]	Performance measure captures the number of forensic profiles that were uploaded to CODIS from cases that grantees analyzed using DNA CEBR grant funds.	2011-2017	CEBR grantee performance reports (aggregated)	Lab activities funded by CEBR grants
Percentage Increase in Forensic DNA Profiles Uploaded to CODIS from the Previous Year	Performance measure captures the percentage increase in the total number of forensic DNA profiles uploaded to NDIS from the previous calendar year.[d]	2014-2017	FBI website hosting data from its National DNA Index System	All lab activities
Offender sample DNA testing				
Number of Convicted Offender and/or Arrestee Database Samples Analyzed[e]	Performance measure captures the number of convicted offender and/or arrestee DNA database samples analyzed using DNA CEBR grant funds.	2011-2017	CEBR grantee performance reports (aggregated)	Lab activities funded by CEBR grants
Number of Convicted Offender and/or Arrestee Database Profiles Uploaded to CODIS	Performance measure captures the number of convicted offender and/or arrestee DNA database profiles that are uploaded to CODIS from samples using DNA CEBR grant funds.	2011-2017	CEBR grantee performance reports (aggregated)	Lab activities funded by CEBR grants
Crime scene DNA analysis and offender sample testing				
Number of CODIS Hits[f]	This performance measure captures the number of CODIS hits made from forensic or convicted offender and/or arrestee DNA database profiles that are uploaded to CODIS using DNA grant funds.	2011-2017	CEBR grantee performance reports (aggregated)	Lab activities funded by CEBR grants

Source: GAO analysis of information provided by the National Institute of Justice | GAO-19-216.

[a] Forensic cases refers to requests for DNA analysis on evidence from crime scenes, victims, and suspects.

[b] A DNA profile contains the genetic constitution of an individual at defined locations (also known as loci) in the DNA. Each person (except identical twins) has a unique DNA profile when used in the context of the national level of the FBI's Combined DNA Index System (CODIS), which now evaluates 20 specific DNA locations.

[c] CODIS is a system that allows federal, state, and local labs to exchange and compare DNA profiles electronically. As a result of processing evidence from crime scenes, only DNA profiles believed to belong to an unknown potential perpetrator can be uploaded into CODIS.

[d] NDIS is one part of CODIS—the national level—containing the DNA profiles contributed by federal, state, and local participating forensic labs. NDIS was implemented in October 1998. All 50 states, the District of Columbia, the federal government, the U.S. Army Criminal Investigation Lab, and Puerto Rico participate in NDIS.

[e] "Convicted offender/arrestee database samples" to refer to DNA analysis on samples taken from convicted offenders, arrestees, and other categories of persons as authorized by law.

[f] The FBI defines a hit as "A confirmed [DNA profile] match that aids an investigation and one or more of the case(s) involved in the match are unsolved."

APPENDIX VII: SELECTED ATTRIBUTES OF SUCCESSFUL PERFORMANCE MEASURES AND CONSEQUENCES IF NOT MET

Table 13 is a summary of selected attributes of successful performance measures, including the potentially adverse consequences if they are missing, that we identified in prior work.

Table 13. Summary of Selected Attributes of Successful Performance Measures and Adverse Consequences of Not Meeting Attribute

Attribute of successful performance measure	Potential adverse consequence of not meeting attribute
Linkage Measure is aligned with program goals.	Behaviors and incentives created by measures do not support program goals.
Clarity Measure is clearly stated and the name and definition are consistent with the methodology used to calculate it.	Data could be confusing and misleading to users.
Measurable target Measure has a numerical goal.	Inability to determine whether performance is meeting expectations.
Core program activities As a group, measures cover the activities that an entity is expected to perform to support the intent of the program.	Not enough information available in core program areas for managers and stakeholders to make key decisions.
Balance As a group, measures ensure that key program priorities are covered.	Lack of balance could create skewed incentives when measures overemphasize some goals.

Source: GAO. | GAO-19-216.

Appendix VIII: DNA Capacity Enhancement and Backlog Reduction Program Permissible Uses of Funds

Table 14. DNA Capacity Enhancement and Backlog Reduction Program (CEBR) Permissible Uses of Funds in 2018 Grant Solicitation

Permissible use of CEBR funds	Description
1. Salary and benefits of lab employees	Funds for additional full-time or part-time employees to directly process, record, screen, and/or analyze forensic DNA and/or DNA database samples. Funds may also be used to hire additional staff to directly perform capacity enhancement-specific activities, such as validating new DNA analysis technologies for the forensic DNA or DNA database lab. Funds are subject to applicable restrictions on supplanting.
2. Overtime for lab staff	Overtime for lab employees (excluding executive personnel) to directly process, record, screen, and/or analyze forensic DNA and/or DNA database samples. Funds may also be used to pay overtime for existing lab employees to directly perform capacity enhancement-specific activities, such as validating new DNA analysis technologies for the forensic DNA or DNA database lab.
3. Training	Funds for training (1) directly related to DNA lab, (2) appropriate continuing and professional education associated with professional meetings and conferences (no more than 8 percent of the award) and (3) travel expenses, registration fees, and learning aids not associated with professional meetings and conferences (no more than 8 percent of the award).
4. Travel	Funds for travel to conduct required site visits to public or private accredited labs that will be conducting DNA analyses on behalf of the eligible state or unit of local government to review procedures and practices prior to initial sample shipment; funds may also be used to make one additional unannounced site visit.
5. Equipment	Funds to upgrade, replace, or purchase lab equipment, instrumentation, and associated computer hardware for the lab.
6. Supplies	Lab supplies for (a) validation, (b) DNA sample analysis and (c) collection kits for offender/arrestee samples.
7. Contracts	Contracts for (a) outsourcing sample processing, (b) DNA audits, (c) process mapping or efficiency studies, (d) warranty, service, or maintenance contracts for equipment, (e) temporary lab employees, (f) validation studies for new DNA analysis technologies and (g) in-house training.
8. Direct administrative expenses	Direct administrative expenses for grant management, not to exceed 3 percent of the federal portion.
9. Costs associated with and including accreditation	Costs associated with accreditation
10. Software	Software expenses associated with running the DNA lab
11. Lab Information Management System (LIMS)	Funds may be used to support existing LIMS and existing LIMS accessories

Source: GAO analysis of CEBR grant program documentation. | GAO-19-216.

Appendix IX: Controls Related to Transparency in Grantee Procurement for DNA Capacity Enhancement and Backlog Reduction Grants

Contracting by grantees and subgrantees under DNA Capacity Enhancement and Backlog Reduction (CEBR) awards is governed by 2 C.F.R. pt. 200 "Uniform Administrative Requirements, Cost Principles, and Audit Requirements for Federal Awards."

Table 15. Office of Justice Programs (OJP) Controls Related to Transparency in Grantee Procurement

Select requirements for OJP and Capacity Enhancement and Backlog Reduction (CEBR) program grantees	OJP control designed to achieve its objectives related to compliance with requirements
OJP review of procurements[a]	
As part of the grant application process, grantees must provide information about proposed procurement contracts.[b] Specifically, grantees should provide the name of the contractor (if known), the amount of the contract, the service to be performed or purchase to be made, and a detailed narrative justification for each contract.	OJP has designed a process to review all budget documentation from grant applicants, including proposed procurement contracts. Specifically, upon receipt of the grant application, the OJP grant manager reviews the documentation to ensure that the proposed budget equals the proposed award amount, and that all proposed costs are allowable, necessary, and applicable, and forwards the documents to a senior grant manager for approval. OJP's Office of the Chief Financial Officer (OCFO) then conducts a second review to ensure that the proposed budget equals the proposed award amount. Grantees cannot move forward with proposed procurement contracts until OJP provides approval.
Grantees must also provide information to OJP during the award period for proposed new contracts or significant changes to existing contracts.	OJP has also designed a process to review proposed new contracts or significant modifications to existing contracts during the award period. OJP grant managers review the requests to determine if the requested changes are allowable under the grant terms and conditions, are necessary and reasonable, and are consistent with the goals and objectives of the grant program. Then, grant managers' supervisors and the National Institute of Justice's Director of the Office of Grant Management review the request. OJP also has grant monitoring checklists for grant managers to use when conducting grant monitoring activities, such as site visits. The checklists include sections for grant managers to review documentation related to contracts that were modified. The grant manager ensures that OJP provided the necessary approvals. Grant managers must certify that they completed all checklist items.

Table 15. (Continued)

Select requirements for OJP and Capacity Enhancement and Backlog Reduction (CEBR) program grantees	**OJP control designed to achieve its objectives related to compliance with requirements**
OJP review of procurements[a]	
Grant recipients are required to submit quarterly financial reports to show obligations and expenditures for contracts.	OJP grant managers review grant recipients' financial reports in conjunction with grant recipients' progress reports to compare the rate of expenditures with the project activity level noted in the progress report and to identify potential problems with the reports. Then, officials from OJP's OCFO perform a financial review to assess and resolve potential problems with the financial report. If OJP notifies a grant recipient of a problem with its financial report but the recipient does not submit a revised financial report, OJP may withhold grant funds.
An awarding agency may conduct pre-procurement review of an applicant's procurement documents when the procurement is expected to exceed the Simplified Acquisition Threshold[c] (currently $250,000 for grantees) and is to be awarded without competition, or if a proposed contract modification changes the scope of a contract or increases the contract amount by more than the Simplified Acquisition Threshold, among other reasons.	When a proposed sole-sourced contract exceeds the Simplified Acquisition Threshold, or when a contract modification changes the contract to exceed the Simplified Acquisition Threshold, grant mangers review the requests against criteria in 2 C.F.R. §200.320(f).[d] Senior grant managers and the National Institute of Justice's Director of the Office of Grants Management provide additional layers of review. Financial analysts in OJP's OCFO then review the procurement information and provide final approval of the non-competitive approach to the procurement. OJP also has grant monitoring checklists for grant managers to use when conducting grant monitoring activities, such as site visits. The checklists include sections for grant managers to review documentation related to contracts that exceed the Simplified Acquisition Threshold and contracts that were modified. The grant manager ensures that OJP provided the necessary approvals. Grant managers must certify that they completed all checklist items.
Grantees are responsible for monitoring subgrantee or contractor activities under federal awards and must assure compliance with applicable federal requirements and performance expectations.	OJP has grant monitoring checklists for grant managers to use when conducting grant monitoring activities, such as site visits. The checklists include a section for grant managers to ask grantees about their monitoring mechanisms for subgrantees and contractors and to review documentation related to such mechanisms, including documentation that demonstrates the grantee is actually conducting monitoring activities for its subgrantees or contractors. Grant managers must certify that they completed all checklist items.
Single audits[e]	
OJP is responsible for following up on single audit findings to provide reasonable assurance that grantees take timely and appropriate action to correct any deficiencies.	The Department of Justice (DOJ) has designed a notification system that helps to ensure that OJP is aware of and responds to findings related to procurements that are discovered during single audits. The Federal Audit Clearinghouse notifies the DOJ Office of the Inspector General (OIG) of significant single audit findings related to procurements. The OIG then notifies OJP, which follows-up with grantees, as applicable.

Select requirements for OJP and Capacity Enhancement and Backlog Reduction (CEBR) program grantees	OJP control designed to achieve its objectives related to compliance with requirements
Reporting to Congress[f]	
Beginning October 2018 and biannually thereafter, for each recipient of funds from the appropriation for "a DNA analysis and capacity enhancement program and for other local, State, and Federal forensic activities"—which includes CEBR grantees—the Attorney General is to report to Senate and House Judiciary Committees the percentage of the amounts that were paid to private laboratories.	OJP added a requirement to the fiscal year 2018 CEBR solicitation for grantees to report the amounts expended under CEBR awards on contracts for DNA analysis. Specifically, the language says: "Recipients will be expected to report amounts expended under the award on contracts to private accredited DNA laboratories for analysis of forensic DNA casework samples or DNA database samples as detailed in the award terms and conditions. Future awards and fund drawdowns may be withheld if reports are delinquent. (In appropriate cases, OJP may require additional reports.)" OJP officials stated they added a special condition to the grant terms and conditions to require CEBR recipients to acknowledge grant funds will be withheld if they do not meet their reporting requirements. OJP officials said they would likely report this information to Congress for each grantee (as opposed to an aggregated amount for all grantees). However, they said they had not finalized the details.

Source: GAO analysis of grantee procurement requirements and OJP information. | GAO-19-216.

[a] The first three requirements come from the CEBR grant solicitation and/or the DOJ Grants Financial Guide. In addition to agreeing to comply with applicable statutes and regulations, CEBR recipients agree to comply with the terms specified in the DOJ Grants Financial Guide and the solicitation. OJP outlines these requirements in the grant terms and conditions. The requirements in the guide and the solicitation are also important elements of transparency in administering CEBR grants. Other requirements come from 2 C.F.R. pt. 200, "Uniform Administrative Requirements, Cost Principles, and Audit Requirements for Federal Awards."

[b] A contract is a legal instrument by which a non-federal entity purchases property or services needed to carry out the project or program under a federal award. A contract is separate from a subaward, which is an award provided by a pass-through entity to a subrecipient for the subrecipient to carry out part of a federal award received by the pass-through entity. 2 C.F.R. §§ 200.22, 200.92.

[c] The "simplified acquisition threshold"—currently set for grantees at $250,000—means the dollar amount below which a non-federal entity may purchase property or services using small purchase methods. 2 C.F.R. § 200.88. Small purchase procedures are those relatively simple and informal procurement methods for securing services, supplies, or other property.

[d] Under 2 C.F.R. § 200.320(f), procurement by noncompetitive proposals may be used only when the item is available only from a single source; the public exigency or emergency for the requirement will not permit a delay resulting from competitive solicitation; the federal awarding agency or pass-through entity expressly authorizes noncompetitive proposals in response to a written request from the non-federal entity; or after solicitation of a number sources, competition is determined inadequate.

[e] Grantees that expend a total of $750,000 or more in federal awards in a fiscal year are required to undergo a single audit of its financial statements and federal awards or a program-specific audit, for the fiscal year. Requirements come from the Single Audit Act Amendments of 1996, Pub. L. No. 104- 156, 110 Stat. 1396 (codified as amended at 31 U.S.C. §§ 7501-7506).

[f] Requirements come from Justice for All Reauthorization Act, Pub. L. No. 114-324, § 3(b)(2)(C), 130 Stat. 1948, 1950.

Per the regulation, grantees that are states (including their agencies and instrumentalities thereof) must follow the same policies and procedures they use for procurements from their non-federal funds. All other grantees and subgrantees, including subrecipients of grantees that are states, must follow the procurement standards set forth in the regulation. Many CEBR grantees are states; nevertheless, Office of Justice Programs (OJP) officials said they apply controls related to transparency in grantee procurement to state grantees and to non-state grantees in the same manner. Table 15 provides a description of selected federal requirements related to transparency in grantee procurement, as well as OJP controls designed to achieve its objectives related to compliance with these requirements.

APPENDIX X: CONTROLS RELATED TO CONFLICTS OF INTEREST FOR DNA CAPACITY ENHANCEMENT AND BACKLOG REDUCTION GRANTS

Criminal conflicts of interest statutes governing Office of Justice Program (OJP) employees who administer DNA Capacity Enhancement and Backlog Reduction (CEBR) grants are codified at 18 U.S.C. Chapter 11. Although these statutes cover a variety of topics related to conflicts of interest, our review focuses on the participation of OJP employees in government actions that may conflict with their personal financial interests, as specifically provided in 18 U.S.C. § 208. We focus on acts affecting personal financial interests because related regulatory requirements cover a broad range of issues directly applicable to OJP employees who administer CEBR grants. Table 16 provides a description of selected federal requirements related to conflicts of interest for federal employees, as well as OJP controls designed to achieve its objectives related to compliance with these requirements. Federal requirements related to the disclosure of conflicts of interest by grantees are found in 2 C.F.R. pt. 200 "Uniform Administrative Requirements, Cost Principles, and Audit Requirements for Federal Awards." These requirements apply to OJP as they oversee grantees, and to CEBR grantees as they interact with OJP and administer grant funding through subgrants or contracts.[1]

Table 16. Office of Justice Programs (OJP) Controls Related to Conflicts of Interest for Federal Employees

Select requirements for OJP and OJP employees	OJP control designed to achieve its objectives related to compliance with requirements
OJP employee conflict of interest requirements[a]	
Unless employees divest their financial interests or obtain a waiver or exemption, they must disqualify themselves from participating in particular matters in which they, or any person whose interests are imputed to them, have a financial interest. Employees should not participate in matters involving specific parties where it is likely to affect the financial interests of a member of their household, a covered party is involved, or under other such circumstances that would cause a reasonable person to question their impartiality.	The Department of Justice (DOJ) has summarized applicable requirements on its ethics website and in the DOJ Ethics Handbook for On and Off-Duty Conduct (Handbook). The Handbook directs employees who may have a conflict of interest or believe their impartiality might be questioned to either disqualify themselves from taking action that could affect their interest, or see OJP's Deputy Designated Agency Ethics Official (DAEO) about authorized alternatives, such as obtaining a waiver or divesting their financial interest. Further, OJP has a web-based tracking system for employees involved in administering Capacity Enhancement and Backlog Reduction Program (CEBR) grants to certify that they do not have conflicts of interest. During each application cycle, employees use the web-based system to (1) review the solicitation and relevant conflicts of interest guidance, (2) review the organizations that submitted applications, and (3) certify that they do not have any conflicts of interest. Employees' supervisors are notified of these certifications. If an employee reports the appearance of a conflict of interest and the supervisor determines a conflict exists, the supervisor may require the staff member to recuse himself or herself from dealing with a specific application or from participating in the entire application cycle.
Employees must not engage in outside employment or other outside activities that conflict with their official duties or employment that involves certain legal practice and matters, including grants.	DOJ has summarized applicable requirements on its ethics website and in the Handbook. Employees are required to obtain written approval from the Deputy Attorney General for OJP to waive these prohibitions.
OJP ethics training requirements[b]	
Agencies must carry out a government ethics education program to teach employees how to identify government ethics issues and obtain assistance in complying with government ethics laws and regulations.	The DOJ has an agency-wide ethics program, which is administered by DOJ's Justice Management Division under the direction of DOJ's DAEO. The DOJ DAEO assigned the Deputy DAEO in OJP to oversee OJP's agency ethics program, as noted above.

[1] A contract is a legal instrument by which a non-federal entity purchases property or services needed to carry out the project or program under a federal award. A contract is separate from a subaward, which is an award provided by a pass-through entity to a subrecipient for the subrecipient to carry out part of a federal award received by the pass-through entity. 2 C.F.R. §§ 200.22, 200.92.

Table 16. (Continued)

Select requirements for OJP and OJP employees	OJP control designed to achieve its objectives related to compliance with requirements
Each new employee must complete initial ethics training.	Each new OJP employee receives new employee ethics training and a copy of the Handbook. At this training, employees are also made aware of the DOJ ethics website.
Those required to file financial disclosure forms must complete annual ethics training.	OJP requires all employees who file public or confidential financial disclosure reports to attend OJP ethics training annually. OJP assigns employees to annual ethics training sessions and makes available a copy of the Handbook to employees at training sessions.
OJP annual financial disclosure requirements[c]	
Senior-level employees must file public financial disclosure reports within 30 days of entering a covered position, and annually thereafter; disclosures must be reviewed and certified by a designated agency ethics offical.	OJP employees in these positions are notified of their obligation to file reports via email. OJP's Ethics Office collects the reports and reviews them to identify the appearance of or actual conflicts of interest within 60 days of receipt. OJP's Ethics Office signs the reports if they believe they are complete and disclose no conflicts of interest. If the OJP Ethics Office finds a conflict, the Deputy DAEO for OJP imposes a remedy, such as disqualification (not participating in the matter), divesture (selling off interests or investments), or waiver (a waiver of the prohibition must be approved by the agency head). Then, OJP's Assistant Attorney General reviews and signs the reports. The Deputy DAEO for OJP notifies DOJ's Departmental Ethics Office when all reports have been filed, reviewed, and signed. DOJ's Departmental Ethics Office then forwards all public financial disclosure reports to the U.S. Office of Government Ethics.
Specified less senior employees must file confidential financial disclosure reports within 30 days of entering a covered position; and annually thereafter; and disclosures must be reviewed and certified by a designated agency ethics official.	OJP employees in these positions are notified of their obligation to file reports via email. Employees' supervisors collect the reports and review them to identify the appearance of or actual conflicts of interest within 30 days of receipt. Employees' supervisors sign the reports if they believe they are complete and disclose no conflicts of interest. If the supervisor finds a conflict, the supervisor and the Deputy DAEO for OJP impose a remedy, such as disqualification, divesture, or waiver.
	The Deputy DAEO for OJP notifies DOJ's Departmental Ethics Office when all reports have been filed, reviewed, and signed. Despite the process described, OJP does not have documentation designating which officials are authorized to certify (provide final signature) these reports or the levels of review required. Further, the designation of reviewers and certifiers within OJP is unclear.
OJP annual financial disclosure requirements[c]	
	OJP officials said that OJP's Deputy DAEO and office heads of OJP sub-components (or deputy office heads) have a role in the review process, but officials did not consistently describe those roles.

Source: GAO analysis of conflicts of interest requirements and OJP information. | GAO-19-216.

[a] Requirements come from 5. C.F.R. pt. 2635, "Standards of Ethical Conduct for Employees of the Executive Branch;" 5 C.F.R. pt. 3801, "Supplemental Standards of Ethical Conduct for Employees of the Department of Justice."

[b] Requirements come from 5 C.F.R. pt. 2638, "Executive Branch Ethics Program"

[c] Requirements come from 5 U.S.C. app. §§ 101-111, "Financial Disclosure Requirements of Federal Personnel" and 5 C.F.R. pt. 2634, "Executive Branch Financial Disclosure, Qualified Trusts, and Certificates of Divestiture. "Senior-level" employees include, among others, employees whose positions are classified above GS-15; or employees whose rate of basic pay is fixed, other than under the General Schedule, at a rate equal to or greater than 120% of the minimum rate of basic pay for a GS-15. "Specified less senior employees" include, among others, employees who occupy positions classified at GS-15 or below or employees whose rate of basic pay is fixed, other than under the General Schedule, at a rate less than 120% of the minimum rate of basic pay for a GS-15. Regarding the administration of CEBR grants, officials said that the Principal Deputy Assistant Attorney General for OJP, National Institute of Justice Director and Principal Deputy Director positions are subject to public disclosure, and all other positions that administer CEBR grants are subject to confidential disclosure.

Table 17. Office of Justice Programs (OJP) Controls Related to Conflicts of Interest for Federal Awarding Agencies and Grantees

Select requirements for OJP and grantees[a]	**OJP control designed to achieve its objectives related to compliance with requirements**
Awarding agencies must establish conflicts of interest policies for grants.	OJP requires grantees to agree during the application process to (1) disclose conflicts of interest in writing to OJP, (2) maintain written standards of conduct covering conflicts of interest and employees participating in the selection, award, and administration of contracts for grants, and (3) have a documented process to check for organizational conflicts of interest with potential contractors.[b] As a condition of award acceptance, grantees must (through execution of the grant award document) agree to comply with (among other award terms and conditions) the Department of Justice's Grants Financial Guide, available on its website, which provides additional guidance concerning conflicts of interest. If a grantee does not agree to these requirements, it will not receive the award. OJP also has grant monitoring checklists for grant managers to use when conducting grant monitoring activities, such as site visits. One of the policies grant managers are instructed to check for when reviewing requirements for procurement policies and procedures is a conflicts of interest policy. Grant managers must certify that they completed all checklist items.
Select requirements for OJP and grantees[a]	OJP control designed to achieve its objectives related to compliance with requirements
Grantees must disclose in writing any conflicts of interest to the awarding agency.	Same control as above.

Source: GAO analysis of conflicts of interest requirements and OJP information. | GAO-19-216.

[a] Requirements come from 2 C.F.R. pt. 200, "Uniform Administrative Requirements, Cost Principles, and Audit Requirements for Federal Awards"

[b] Under these regulations, a contract is a legal instrument by which a non-federal entity purchases property or services needed to carry out the project or program under a federal award. 2 C.F.R. § 200.22.

Table 17 provides a description of selected federal requirements related to conflicts of interest for federal awarding agencies and grantees, as well as OJP controls designed to achieve its objectives related to compliance with these requirements.

APPENDIX XI: CONTROLS RELATED TO LOBBYING FOR DNA CAPACITY ENHANCEMENT AND BACKLOG REDUCTION GRANTS

Federal law prohibits recipients of federal awards from using appropriated funds for lobbying activities in connection with the award, and sets forth several requirements related to lobbying "certification" and "disclosure."[2]

Lobbying certification refers to agreeing not to use appropriated funds to lobby. Lobbying disclosure refers to disclosing lobbying activities with respect to the covered federal action (in this case, DNA Capacity Enhancement and Backlog Reduction (CEBR) program grants) paid for with nonappropriated funds.

Federal regulation requires recipients of all federal awards over $100,000 to file certification documents and disclosure forms (if applicable) with the next tier above. Disclosure forms, but not certification documents, are to be forwarded from tier to tier until received by the federal agency).[3] In the case of CEBR grants, tiers include the Office of Justice Programs (OJP), grantees, subgrantees, contractors under grantees

---

[2] 28 C.F.R. pt. 69, "New Restrictions on Lobbying," implements 31 U.S.C. § 1352, commonly referred to as the "Byrd Amendment." Lobbying in the context of this statute refers to paying any person for influencing or attempting to influence an officer or employee of any agency, a Member of Congress, an officer or employee of Congress, or an employee of a Member of Congress in connection with any covered Federal actions. See also, 2 C.F.R. pt. 200, Appendix II(I) which contains similar requirements. These requirements are different from lobbying registration and disclosure requirements set forth in the Lobbying Disclosure Act of 1995, as amended (codified at 2 U.S.C. §§ 1601 et seq.).

[3] The required certification document is set forth in appendix A to 28 C.F.R. pt. 69. The required disclosure form is set forth in appendix B to 28 C.F.R. pt. 69. Disclosure forms are only required if the recipient has used or plans to use nonappropriated funds to lobby with respect to the award. Additional exceptions to the prohibition and disclosure requirements apply.

and subgrantees, and subcontractors. Table 18 provides a description of selected federal requirements related to lobbying, as well as OJP controls designed to achieve its objectives related to compliance with these requirements.

Table 18. Office of Justice Programs (OJP) Controls Related to Lobbying

Select requirements for Capacity Enhancement and Backlog Reduction (CEBR) grantees and others[a]	**OJP control designed to achieve its objectives related to compliance with requirements**
Grantees are prohibited from using appropriated funds, directly or indirectly, to pay for activities (e.g., advertisements, printed materials, etc.) to influence government officials to support or oppose legislation, policies, or other matters (unless the activity is expressly authorized by Congress).[b]	As a condition of award acceptance, grantees must agree to comply with OJP's "General Conditions," which includes (1) language prohibiting the use of federal funds for lobbying, and (2) a requirement that grantees follow applicable lobbying laws, as set forth in the Department of Justice's Grants Financial Guide. OJP also follows the grantees' use of grant funds through grant monitoring activities. For example, OJP grant managers review grant recipients' financial reports in conjunction with grant recipients' progress reports to compare the rate of expenditures with the project activity level noted in the progress report and to identify potential problems with the reports, among other monitoring steps.
Applicants and recipients of federal awards must file with the agency a certification with specific language from appendix A to 28 C.F.R. pt. 69, that the applicant or grantee has not made and will not make prohibited lobbying payments.	As a condition of award acceptance, grantees must (through execution of the grant award document) agree to the certification document (among other award terms and conditions). When accepting the award, grantees must print, initial each page, and sign all award documents, and then email a scanned copy of the signed documents to the Office of the Chief Financial Officer (OCFO) for review. If a grantee does not agree to these requirements, it will not receive the award.
If the applicant or grantee has made or has agreed to make a payment using nonappropriated funds that would be prohibited if appropriated	OJP requires that all CEBR applicants submit the lobbying disclosure form as part of the grant application process. Upon submission, a grant manager reviews the form for completeness and content and checks a box in an application
funds were used, it must file with the agency a disclosure form as set forth in appendix B to 28 C.F.R. pt. 69.	review checklist. In the event there is lobbying activity disclosed, the grant manager forwards the form to OJP attorneys and alerts them to the lobbying activity for their review.
Certifications and disclosure forms are required to be filed with the tier[c] above for subgrants and contracts greater than $100,000. All disclosure forms, but not certifications, are to be forwarded from tier to tier until received by the agency. Grantees and subrecipients are to disclose subsequent lobbying events that require disclosure or specified events	As stated above, OJP has designed a process to collect certification documents and disclosure forms from CEBR applicants. However, the certification document that CEBR grant applicants agree to does not state in clear terms what the specific requirements of the regulation are or how they are to be carried out. The certification document sets forth the requirements for filing certifications and disclosures, as detailed above. It states: "The Applicant shall require that the

Table 18. (Continued)

Select requirements for Capacity Enhancement and Backlog Reduction (CEBR) grantees and others[a]	OJP control designed to achieve its objectives related to compliance with requirements
that materially affect the accuracy of previously filed disclosures at the end of each quarter.	language of this certification be included in the award documents for all subgrants and procurement contracts (and their subcontracts) funded with Federal award funds and shall ensure that any certifications or lobbying disclosures required of recipients of such subgrants and procurement contracts (or their subcontractors) are made and filed in accordance with 31 U.S.C. § 1352
	" Upon reading this, grantees or subrecipients may still be unsure of their responsibilities under the regulation and when they are to carry them out. For example, they may not understand what mechanism to use to obtain a certification document or that disclosure forms must be forwarded to the tier above until received by OJP.
The agency is to take such actions as are necessary to ensure that the lobbying requirements are vigorously implemented and enforced.[d]	As stated above, OJP has designed a process to collect certification documents and disclosure forms from CEBR applicants. However, it is not taking actions to ensure that grantees are requiring subrecipients to certify and disclose, and that grantees forward disclosures according to the requirements set forth above. For example, grant managers use checklists to guide grant monitoring activities, such as site visits, to ensure that grantees are following the grant requirements they agreed to during the grant application process. While these checklists have instructions for asking questions about other grant requirements, they do not include instructions to ensure grantees are collecting certifications documents and disclosure forms from tiers below them and forwarding disclosure forms to OJP.

Source: GAO analysis of federal lobbying requirements and OJP information. | GAO-19-216.

[a]Unless otherwise noted, requirements come from 28 C.F.R. pt. 69, "New Restrictions on Lobbying" which implements 31 U.S.C § 1352, commonly referred to as the "Byrd Amendment." See also, 2 C.F.R. pt. 200, Appendix II(I) which contains similar requirements. Requirements are applicable to federal awards over $100,000. However, all CEBR grants are for at least $150,000. Additional exceptions and requirements apply.

[b]This requirement is set forth in 18 U.S.C. § 1913.

[c]In the case of CEBR grants, tiers include OJP, grantees, subgrantees, contractors under grantees and subgrantees, and subcontractors.

[d]31 U.S.C. § 1352; see also 28 C.F.R. § 69.410.

APPENDIX XII: COMMENTS FROM THE DEPARTMENT OF JUSTICE

U.S. Department of Justice

Office of Justice Programs

Office of the Assistant Attorney General

Washington, D.C. 20531

MAR - 6 2019

Ms. Gretta L. Goodwin
Director
Homeland Security and Justice
Government Accountability Office
441 G Street, NW
Washington, DC 20548

Dear Ms. Goodwin:

Thank you for the opportunity to review and comment on the draft Government Accountability Office (GAO) report entitled, "DNA Evidence: DOJ Should Improve Performance Measurement and Properly Design Controls for Nationwide Grant Program" (GAO-19-216). In this report, the GAO confirms what the National Institute of Justice (NIJ) has maintained for over a decade – that laboratory capacity continues to increase, but progress to decrease backlogs in some jurisdictions has been tempered by the increased demand from law enforcement and prosecutors to conduct DNA testing. This growth in demand is due, in part, to advancements in the technology and requests to process DNA from unsolved homicides and sexual assaults, potential erroneous convictions, and unidentified human remains.

OJP's NIJ – the research, development, and evaluation arm of the Department of Justice (DOJ) – is the only Federal agency supporting forensic science programs dedicated to research, development, and evaluation, in conjunction with capacity building and technical assistance. As part of its overall strategic plan to meet the needs of the forensic science community, NIJ has aligned its programs with long-term national strategies and short term needs that arise primarily from state and local forensic laboratories that face the dual challenges of satisfying the increasing demand for forensic testing, while keeping up with the latest advances and quality-related requirements. Many forensic laboratories continue to require significant improvements to their infrastructure, relating not only to DNA processing, but other forensic disciplines that are critical in violent crime cases, especially when they lack DNA evidence.

NIJ remains steadfast that the quality of DNA testing should always remain a priority for our nation's forensic laboratories, while the two overarching goals of the DNA CEBR Grant Program are for DNA laboratories to: (1) enhance their capacity through increased productivity and increased efficiency, and (2) decrease their backlogs of DNA samples awaiting analysis.

The draft GAO report contains four Recommendations for Executive Action, directed to the Principal Deputy Assistant Attorney General of OJP. For ease of review, the recommendations are restated in bold text below, and are followed by OJP's response.

1. **The Principal Deputy Assistant Attorney General for OJP should consistently document CEBR program-wide goals to clarify intended program results.**

   The Office of Justice Programs agrees with this Recommendation for Executive Action. OJP has worked to ensure that the two overarching goals of the DNA CEBR Program - (1) enhance their capacity through increased productivity and increased efficiency, and (2) decrease their backlogs of DNA samples awaiting analysis – have repeatedly and consistently been communicated to applicants and grant recipients through annual solicitations, webinars, seminars, and supporting grant documentation. OJP has also communicated through reports and publications, accessible to the forensic field and public, that DNA CEBR program funds can be used for improvements in quality and throughput that are achieved through innovations, such as more sensitive chemistries, faster technologies, and streamlined workflows.

   OJP will further review how DNA CEBR program-wide goals are communicated to grant recipients, Congress, and stakeholders to determine what changes may be needed to ensure that the goals are consistently communicated.

2. **The Principal Deputy Assistant Attorney General for OJP should ensure that performance measures for each CEBR program-wide goal fully reflect appropriate attributes of successful performance measures.**

   The Office of Justice Programs agrees with this Recommendation for Executive Action. NIJ captures data for multiple performance measures that are intended to reflect the comprehensive value and impact of the DNA CEBR Grant Program, and three outcome metrics, including measures that are aligned with the two overarching goals (e.g., number of cases backlogged, number of samples analyzed per analyst per month, and turnaround time) and metrics that directly measure the impact of these funds (e.g., number of cases analyzed, and number of CODIS hits). In addition, annually since 2015, OJP has reported the increase in the number of DNA profiles uploaded to CODIS.

   To better describe the interrelated nature of the measures and how each measure contributes to a comprehensive picture of program impact, OJP will develop a logic model for the DNA CEBR Grant Program. This logic model will more clearly indicate for grant recipients the measures and quantitative goals for their project and organization that may be achieved with support of funding from the DNA CEBR Program.

   OJP will also evaluate whether there is an additional suitable target value that can be accurately reported, in compliance with the Government Performance and Results Act.

3. **The Principal Deputy Assistant Attorney General for OJP should document which employee positions have been delegated certification (final signature) authority for confidential disclosure reports and specify required levels of review and approval.**

The Office of Justice Programs agrees with this Recommendation for Executive Action. OJP will adopt its own version of the policy implemented by DOJ's Justice Management Division (JMD), which OJP has been following as a practical matter. The JMD policy can be found at https://www.justice.gov/jmd/procedures-confidential-financial-disclosure-system.

4. **The Principal Deputy Assistant Attorney General should: (1) clarify its guidance to grantees to specify what their requirements are under 28 C.F.R. pt 69 with regard to obtaining lobbying certification documents, and obtaining and forwarding to OJP lobbying disclosure forms, from tiers beneath them, and (2) design a control to follow-up with grantees to help ensure they are meeting these requirements.**

The Office of Justice Programs agrees with this Recommendation for Executive Action. In February 2019, as part of the implementation of centralized standard certifications, a lobbying certification was added to the System for Award Management (SAM.gov), which specifies that the applicant certify they will comply with the requirements of the Lobbying Disclosure Act of 1995. All award applicants are required to register with SAM.gov and, as such, will provide their certification of compliance at the point of registration.

OJP provides guidance to grant recipients regarding lobbying certification and disclosure forms in OJP's grant solicitations and the DOJ's Grants Financial Guide. Beginning with the Fiscal Year 2019 grant awards, OJP will update its annual desk review checklist to follow-up with grant recipients to ensure any applicable disclosure forms are collected and submitted to OJP, based on the requirements in 28 CFR 69.110.

Sincerely,

Matt M. Dummermuth
Principal Deputy Assistant Attorney General

cc: Lee Lofthus
Assistant Attorney General for Administration
U.S. Department of Justice

Maureen A. Henneberg
Deputy Assistant Attorney General

cc: David Muhlhausen
Director
National Institute of Justice

Rafael A. Madan
General Counsel

Ralph E. Martin
Director
Office of Audit, Assessment, and Management

Leigh Benda
Chief Financial Officer
Officer of the Chief Financial Officer

Robert Davis
Acting Director
Office of Communications

Richard P. Theis
Director, Audit Liaison Group
Internal Review and Evaluation Office
Justice Management Division

OJP Executive Secretariat
Control Title IT20190131102353

In: DNA
Editor: Tomáš Koláček

ISBN: 978-1-53616-117-5

Chapter 5

DNA EVIDENCE: PRELIMINARY OBSERVATIONS ON DOJ'S DNA CAPACITY ENHANCEMENT AND BACKLOG REDUCTION GRANT PROGRAM*

Gretta L. Goodwin

WHY GAO DID THIS STUDY

Many state and local crime labs have backlogs of requests for DNA analysis of crime scene evidence, as reported by grantees participating in DOJ's CEBR grant program. These backlogs can include sexual assault kits. Since 2011, DOJ's Office of Justice Programs—the primary grant-making arm of DOJ—has awarded nearly $500 million to states and local jurisdictions through the CEBR grant program to help reduce DNA evidence awaiting analysis at crime labs. There have been concerns that these backlogs of unanalyzed evidence have enabled serial offenders to reoffend or have delayed justice.

* This is an edited, reformatted and augmented version of United States Government Accountability Office, Testimony, Before the Committee on the Judiciary, U.S. Senate, Publication No. GAO-18-651T, dated July 18, 2018.

This statement is based on preliminary observations and analyses from GAO's ongoing review of (1) the level of crime scene DNA evidence backlogs among CEBR grantees and the factors that contribute to such backlogs; (2) the extent to which DOJ has clearly defined goals for CEBR; and (3) the extent to which OJP has controls for CEBR related to federal conflicts of interest and lobbying requirements. To develop these preliminary findings, GAO reviewed CEBR grantee data from 2011-2016 (the latest data available) and studies relevant to the DNA backlog, visited selected labs, and interviewed DOJ officials, among others.

What GAO Recommends

GAO is not making recommendations in this testimony but will consider them, as appropriate, as it finalizes its work.

What GAO Found

GAO's preliminary analysis found that, among the Department of Justice's (DOJ) DNA Capacity Enhancement and Backlog Reduction Program (CEBR) grantees (state and local entities with forensic crime labs), the reported aggregated backlog of crime scene DNA analysis requests has increased by 77 percent from 2011-2016. The growth in this reported aggregate backlog is the result of labs receiving more requests than they were able to complete, although they were receiving and completing more requests, as shown in the Figure below.

GAO's preliminary analysis also found that the National Institute of Justice (NIJ)—the component within DOJ's Office of Justice Programs (OJP) that is responsible for administering CEBR grants—has not defined CEBR program-wide goals in clear, specific, and measurable terms. Additionally, GAO's ongoing work identified statements in NIJ and CEBR program documentation that communicated program-wide goals, but the documentation did not consistently identify the same goals or cite the same number of goals. GAO continues to evaluate CEBR program goals and is in the process of evaluating related CEBR performance measures as part of its ongoing work.

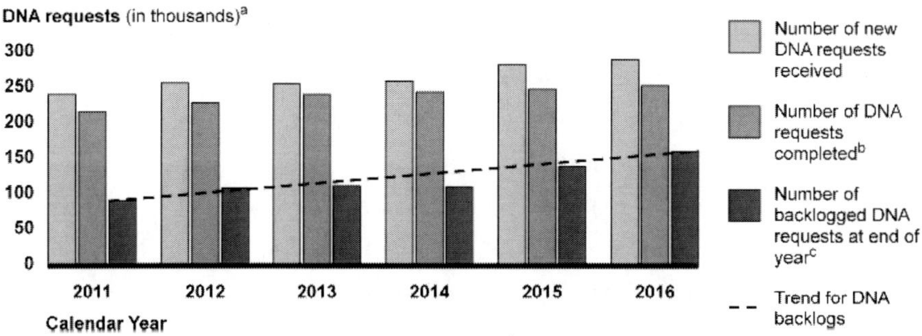

Source: GAO analysis of CEBR grant program. | GAO-18-651T.

[a]By "DNA requests" we mean requests for biology screening and/or DNA testing such that if a request requires one or both, it is counted once.

[b]Some requests may be closed by administrative means rather than through analysis, such as when a suspect pleads guilty before the evidence is analyzed or when the victim declines to press charges. These requests are included in the number of requests received but are not included in the number of requests completed.

[c]The National Institute of Justice defines a "backlogged" request for crime scene DNA analysis as a request that has not been completed within 30 days of receipt in the laboratory. In the figure above, backlogs are bigger than the difference between the number of new requests and the number of requests completed because backlogs also include outstanding requests from previous years.

Preliminary Observations of Levels of Crime Scene DNA Analysis among DNA Capacity Enhancement and Backlog Reduction Grant Program Grantees.

GAO's preliminary analysis found that OJP has some controls to implement federal requirements associated with conflicts of interest and lobbying that apply to both OJP CEBR grant administrators as well as recipients of CEBR grant funding, but OJP has not fully established all appropriate controls related to lobbying.

Chairman Grassley, Ranking Member Feinstein, and Members of the Committee:

Thank you for the opportunity to discuss preliminary observations from our ongoing work on backlogs of DNA evidence.[1] In the criminal justice context, DNA analysis may be used to develop investigative leads, link an offender or victim to a crime scene, or confirm or disprove an account of a crime. Many state and local crime labs have backlogs of requests for DNA analysis of crime scene evidence, as reported by grantees participating in the Department of Justice's (DOJ) DNA Capacity Enhancement and Backlog Reduction grant program (CEBR).[2] These backlogs can include sexual assault kits (SAKs).[3] There have been concerns that these backlogs of unanalyzed evidence have enabled serial offenders to reoffend or have delayed justice. In addition, questions have been raised about potential improper connections among those who profit from DNA testing and those who advocate for CEBR funding, such as private companies providing financial assistance to CEBR funding advocates.

Since 2011, DOJ has awarded nearly $500 million to states and local jurisdictions through CEBR to help reduce DNA evidence awaiting analysis at crime labs. CEBR funds can be used to process DNA evidence collected from crime scenes or offenders.

My statement today is based on preliminary observations and analyses from our ongoing review for Senator Grassley of DNA evidence backlogs and the CEBR grant program. Specifically, I will address the following topics: (1) the level of crime scene DNA evidence backlogs among CEBR grantees and the factors that contribute to such backlogs;[4] (2) the extent to

---

[1] DNA stands for deoxyribonucleic acid and is the genetic material present in the nucleus of cells.

[2] According to DOJ, a backlogged request for analysis of crime scene evidence is a request that has not been completed within 30 days of receipt in the laboratory.

[3] According to DOJ, a SAK is a package of materials used to collect and store biological evidence from a victim's or suspect's body by a medical professional in cases of alleged sexual assault. The contents of a SAK vary by jurisdiction but generally include swabs and collection envelopes for biological material.

[4] As used in this statement, "crime scene DNA evidence" refers to evidence from crime scenes, victims, and suspects collected by law enforcement or other responders after a crime. This differs from DNA samples collected from convicted offenders, arrestees, and others pursuant to law.

which DOJ has clearly defined goals for CEBR; and (3) the extent to which DOJ's Office of Justice Programs (OJP)—the primary grant-making arm of DOJ—has controls for CEBR related to federal conflicts of interest and lobbying requirements.

As part of our ongoing work, we reviewed CEBR grantee data and studies relevant to the DNA backlog, we visited selected labs, and we interviewed DOJ officials, among others. To assess the backlog of DNA evidence among CEBR grantees, we collected yearly baseline performance data from CEBR grantees from 2011 through 2016. We selected 2011 to coincide with the year in which two grant programs were combined into what is now known as the CEBR grant program, and 2016 was the latest year for which grantee data were available. To assess the extent to which DOJ-compiled CEBR baseline data are reliable, we completed a number of data reliability steps, including discussing data entry issues with grantees and running logic tests on the data. Despite some limitations found with regards to validating and reporting requirements, after completing these steps, we determined that CEBR data were sufficiently reliable for the purposes of illustrating year-over-year aggregate trends among CEBR grantees.

Additionally, to identify factors that contribute to backlogs of unanalyzed DNA evidence at laboratories, we reviewed relevant studies and discussed factors with DNA evidence stakeholders, including DOJ officials, CEBR grantees, forensic crime laboratory directors, and relevant academics and practitioners. We summarized information from these sources to identify common factors, and we included illustrative examples of the types of factors we have identified to date in our statement. To assess the extent to which DOJ has clearly defined goals for CEBR, we are reviewing OJP CEBR documentation, including the most recent CEBR grant solicitation and CEBR program performance reports. We also discussed program goals with OJP officials. Finally, to assess the extent to which OJP has controls associated with federal conflicts of interest and lobbying requirements, we identified federal statutes and regulations governing conflicts of interest and lobbying. We are reviewing OJP documentation and statements from OJP officials to determine the extent to which OJP has controls in place consistent with those requirements.

This work is being conducted in accordance with generally accepted government auditing standards. Those standards require that we plan and perform the audit to obtain sufficient, appropriate evidence to provide a reasonable basis for our findings and conclusions based on our audit objectives. We believe that the evidence obtained provides a reasonable basis for our findings and conclusions based on our audit objectives.

BACKGROUND

DNA Analysis of Crime Scene Evidence

Analysis of DNA evidence from crime scenes can help law enforcement link offenders or victims to crime scenes. After crimes occur, law enforcement submits physical evidence from crime scenes, victims, and suspects (hereafter referred to as "crime scene evidence") to labs for analysis. Labs then perform "DNA analysis," which, as used in this statement, refers to (1) biology screening (locating, screening, identifying, and characterizing blood and other biological stains and substances); and/or (2) DNA testing (identifying and comparing DNA profiles in biological samples).[5] In order to compare the victim's or offender's DNA profile to the recovered crime scene DNA, the lab will need to have known biological samples available. Thus, samples are generally collected from victims and may also be collected from others—such as suspects, crime

[5] We use this definition of "DNA analysis" because the CEBR program counts requests that undergo biology screening and/or DNA testing as one request (such that if a request requires one or both, it is counted once). A DNA profile is the genetic constitution of an individual at defined locations (also known as loci) in the DNA. Each person (except identical twins) has a unique DNA profile when used in the context of national level of the FBI's Combined DNA Index System (CODIS), which now evaluates 20 specific DNA locations. We discuss CODIS later in this report.

scene personnel, first responders, and consensual sexual partners (in cases of sexual assault).[6]

Matching DNA Profiles in the FBI's Combined DNA Index System

Matching DNA profiles from unknown potential offenders to existing DNA profiles can help law enforcement develop investigative leads. If a case has no suspects to compare the DNA evidence to, the DNA profile of the unknown potential offender can be entered in the Federal Bureau of Investigation's (FBI) Combined DNA Index System (CODIS), where it can be compared to existing DNA profiles at the local, state, or national level.[7] Labs can then compare unknown potential offender profiles to other profiles already in CODIS, including:

1. Profiles generated from evidence taken from other crime scenes and connected to other unknown potential offenders.
2. Profiles generated from samples taken from known convicted offenders, arrestees, and others as required by law (hereafter "offender samples"). According to DOJ, the federal government, all 50 states, the District of Columbia, and Puerto Rico have laws requiring the collection of DNA samples from individuals convicted of certain crimes; in addition, the federal government, over half of the states, and the District of Columbia have laws

[6] In practice, evidence often contains a mixture of DNA from more than one person. These mixtures can be challenging to analyze and interpret. Additionally, DNA analysis may result in "partial" profiles, which may occur when samples have low quantities of DNA or are exposed to extreme environmental conditions, among other things.

[7] CODIS is the generic term used to describe the FBI's program of support for criminal justice DNA databases as well as the software used to run these databases. CODIS has multiple levels where DNA profiles can be stored and searched; the local level (for city and county DNA laboratories), state level and national level. As a result of processing evidence from crime scenes, only DNA profiles believed to belong to an unknown potential offender can be uploaded into CODIS. Thus, DNA collected from victims, suspects, first responders, and other known persons are processed for comparison and elimination purposes only and are not retained in CODIS.

authorizing the collection of DNA from individuals arrested for certain crimes.[8]

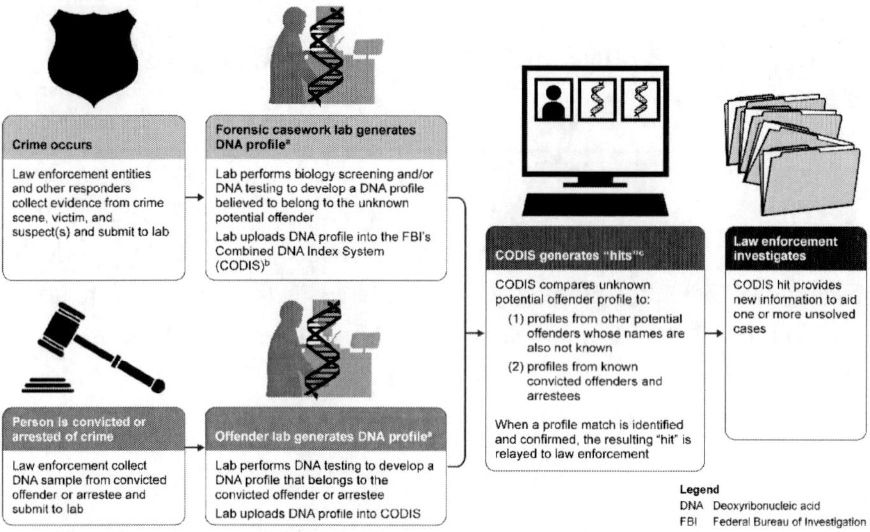

Source: GAO analysis of DOJ information and a stakeholder interview; Art Explosion (clip art). | GAO-18-651T.

[a]We use the term "forensic casework lab" to refer to labs that perform DNA analysis on evidence from crime scenes, victims, and suspects; we use the term "offender lab" to refer to labs that perform DNA analysis on samples taken from convicted offenders, arrestees, and other categories of persons as authorized by law; some labs do both.

[b]CODIS is the generic term used to describe the FBI's program of support for criminal justice DNA databases as well as the software used to run these databases. Only federal, state, or local government labs can participate in CODIS. Government labs may outsource DNA analysis to private labs, but government labs must take or retain ownership of the DNA data for entry into CODIS.

[c]The FBI defines a hit as "a confirmed match that aids an investigation and one or more of the case(s) involved in the match are unsolved."

Figure 1. How DNA Profiles Are Generated and Compared to Help Law Enforcement Develop Investigative Leads.

[8] DNA samples may also be taken from non-U.S. person detainees and other categories of persons, as authorized by law.

When an unknown potential offender's profile matches another profile within CODIS, a "hit" or investigative lead may be developed and shared with law enforcement, as shown in Figure 1 above.[9]

Only federal, state, or local government labs that meet the FBI's Quality Assurance Standards can participate in CODIS.[10] As of January 1, 2018 there were 201 labs that participated in CODIS in the U.S. Of these, 143 performed just forensic casework DNA analysis, 4 performed just offender sample DNA analysis, and 54 performed both. According to the FBI, as of May 2018, the national level of CODIS contained over 16 million profiles generated from offender samples and over 850,000 profiles generated from crime scene evidence.[11] Also, the FBI reported that as of May 2018, CODIS had produced over 422,000 hits that aided more than 406,000 investigations.

The CEBR grant program is administered by the National Institute of Justice (NIJ), a component within OJP. NIJ, the research arm of DOJ, is responsible for evaluating programs and policies that respond to crime, and providing and administering awards for DNA analysis and forensic activities, among other criminal justice activities. The CEBR grant program is funded by an appropriation "for a DNA analysis and capacity enhancement program and for other local, State, and Federal forensic activities."[12] The broad appropriations language enables NIJ to allocate funding for a variety of forensic programs at funding levels established by the agency; however, congressional reports accompanying the appropriation have directed that OJP make funding for DNA analysis and

[9] The FBI defines a hit as "A confirmed match that aids an investigation and one or more of the case(s) involved in the match are unsolved."

[10] FBI, Quality Assurance Standards for Forensic DNA Testing Laboratories (September 1, 2011). FBI, Quality Assurance Standards for DNA Databasing Laboratories (September 1, 2011). Government labs may outsource DNA analysis to private labs, but government labs must take or retain ownership of the DNA data for entry into CODIS.

[11] The national level of CODIS, called the National DNA Index System, contains DNA profiles contributed by federal, state, and local participating forensic labs.

[12] The appropriation language states that funds are "for a DNA analysis and capacity enhancement program and for other local, State, and Federal forensic activities, including the purposes authorized under section 2 of the DNA Analysis Backlog Elimination Act of 2000 (Public Law 106–546) (the Debbie Smith DNA Backlog Grant Program)." The purposes of the CEBR grant program are generally similar to the purposes of the Debbie Smith DNA Backlog Grant Program. There is no additional statutory authorization for the program.

capacity enhancement a priority.[13] CEBR awards can be used to enhance capacity and reduce backlogs at government labs that analyze crime scene DNA evidence and/or process offender DNA samples. NIJ defines a "backlogged" request for analysis of crime scene evidence as a request that has not been completed within 30 days of receipt in the laboratory.

CEBR is a formula grant program that dates back to 2004.[14] Grant awards are made non-competitively to states and units of local government based on a formula set by DOJ that allocates certain amounts to each state. This formula takes into account each state's population and associated crime, and guarantees a minimum amount for eligible applicants from each state. CEBR has broad participation from states and local jurisdictions. For instance, in 2017 OJP awarded $61 million in CEBR grants to 131 grantees in 49 states, the District of Columbia, and Puerto Rico.

Preliminary Analysis of CEBR Data Show That the Backlog of Requests for Crime Scene DNA Analysis Is Increasing and Stakeholders Attribute This to Various Factors

Our preliminary analysis of CEBR grant program data show that the backlog of requests for crime scene DNA analysis has increased by 77 percent from 2011 through 2016, and that demand for such DNA analysis

[13] The Justice for All Reauthorization Act of 2016, Pub. L. No. 114-324, § 3(a), requires that not less than 75 percent of the funds made available under this appropriation be provided for grants for activities described under paragraphs (1), (2), and (3) of section 2(a) of the DNA Analysis Backlog Elimination Act of 2000. Those purposes include (1) To carry out, for inclusion in the Combined DNA Index System of the Federal Bureau of Investigation, DNA analyses of samples collected under applicable legal authority; (2) To carry out, for inclusion in such Combined DNA Index System, DNA analyses of samples from crime scenes, including samples from rape kits, samples from other sexual assault evidence, and samples taken in cases without an identified suspect; and (3) To increase the capacity of laboratories owned by the State or by units of local government to carry out DNA analysis of samples specified in the purposes above.

[14] In 2011, grant programs that separately funded labs that analyzed (1) crime scene DNA evidence, and (2) offender DNA samples, were combined into one grant program called the DNA Backlog Reduction Program. In 2014, this was renamed the DNA Capacity Enhancement and Backlog Reduction Program.

has outpaced laboratory capacity. In our review, we identified numerous factors that have contributed to an increased demand for DNA analysis beyond laboratories' capacities, including scientific advancements in DNA analysis technology and state laws requiring testing of certain DNA evidence.

Preliminary Analysis of CEBR Data Show an Increasing Backlog for Crime Scene DNA Analysis at Laboratories among CEBR Grantees, though Backlogs Vary Among Individual Labs

We found that, among CEBR grantees, the reported aggregated backlog of requests for crime scene DNA analysis has increased by 77 percent from 2011 through 2016. As part of the grant application process, NIJ requires applicants for CEBR grants to provide data from all labs in their jurisdiction, even if certain labs will not be using CEBR funds. NIJ does this to assist in understanding nationwide trends in DNA analysis backlogs.[15] The reported growth in the aggregate backlog among CEBR grantees is the result of labs receiving more requests than they were able to complete over time, as shown in the Figure below.

Although reported aggregate trends show an increase in the backlog among CEBR grantees, the data also reveal that this increase is not uniform across all labs. For example, among the 118 grantees for which we had data from 2011 through 2016, 30 grantees (25 percent) reported an overall decrease in the backlog. In addition, data from CEBR grantees show differences in the average time it takes to process requests (turnaround time) among grantees.

[15] As of January 1, 2018, according to information provided by OJP, CEBR grantees represented 188 of the 197 laboratories that perform forensic casework DNA analysis in the U.S. and participate in CODIS. Although this information is for 2018, NIJ officials said that CEBR grantees have generally been representative of all labs in the U.S. that participated in CODIS in prior years as well. In addition, according to NIJ officials, NIJ collects data from CEBR grantees in years they did not apply for CEBR funding to keep the data consistent. We are continuing to assess the extent to which CEBR data are nationally representative for the years presented as part of our ongoing work.

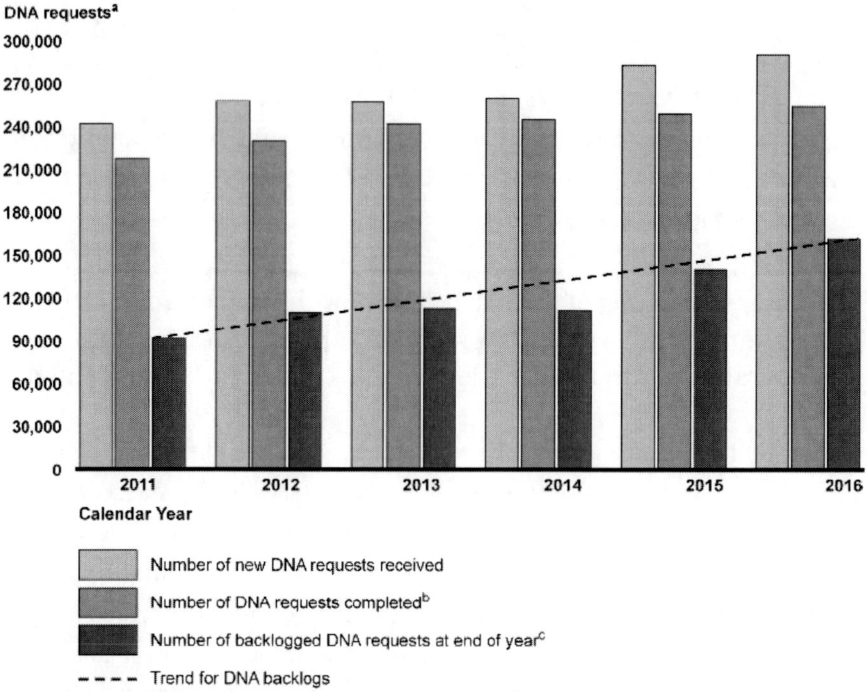

Source: GAO analysis of CEBR grant program. | GAO-18-651T.

[a]By "DNA requests" we mean requests for biology screening and/or DNA testing such that if a request requires one or both, it is counted once.

[b]Some requests may be closed by administrative means rather than through analysis, such as when a suspect pleads guilty before the evidence is analyzed or when the victim declines to press charges. These requests are included in the number of requests received but are not included in the number of requests completed.

[c]The National Institute of Justice defines a "backlogged" request for crime scene DNA analysis as a request that has not been completed within 30 days of receipt in the laboratory. In the figure above, backlogs are bigger than the difference between the number of new requests and the number of requests completed because backlogs also include outstanding requests from previous years.

Figure 2. Preliminary Observations of Requests for Crime Scene DNA Analysis, Requests Completed, and Backlogged Requests among DNA Capacity Enhancement and Backlog Reduction Grant Program Grantees.

Stakeholders also stated, and NIJ has reported, that labs generally have shorter average turnaround times for requests associated with violent crimes than for requests associated with non-violent crimes—because labs generally prioritize requests associated with violent crimes.[16] For our ongoing review, we continue to analyze CEBR data and data from other sources pertaining to this issue.

Various Factors, Such as Scientific Advancements, Have Increased Demand for DNA Analysis beyond Laboratories' Capacities

Based on a review of a selection of studies and discussions with DNA evidence stakeholders, we identified the following factors that are reported to have contributed to an increased demand for crime scene DNA analysis beyond laboratories' capacities. As a result, these factors are believed to have helped contribute to increased backlogs:[17]

Scientific Advancements

Recent scientific advancements have increased the quality of DNA analysis by allowing lab analysts to obtain DNA profiles from smaller amounts of biological evidence. This has increased the amount of evidence that is eligible to be analyzed and, as a result, has increased the demand for DNA testing.

Decreases in Turnaround Times

One DNA evidence stakeholder was able to produce preliminary data demonstrating that, as a general trend, labs that decreased their turnaround time saw corresponding increases in requests from law enforcement. Other DNA stakeholders, including NIJ, made similar observations.[18]

---

[16] DNA evidence stakeholders include DOJ officials, CEBR grantees, forensic crime laboratory directors, and relevant academics and practitioners.

[17] We identified and reviewed 22 government reports and academic journal articles about factors that contribute to backlogs of unanalyzed DNA evidence at labs and inventories of unanalyzed DNA evidence in law enforcement custody. Throughout the course of our audit, we also discussed factors that contribute to backlogs and/or inventories with stakeholders from 15 different entities, including CEBR grantees and the American Society of Crime Lab Directors, among others.

[18] In a market environment, if a price decreases, quantity demanded generally increases. State and local labs are generally funded by state or local appropriations and thus their services are

Increased Awareness among Law Enforcement and the Public

Increased awareness among law enforcement officers of the value of DNA analysis in solving current and older cases has led to law enforcement agencies submitting more DNA evidence to labs for analysis. Further, NIJ and other stakeholder officials we interviewed stated that the volume of DNA profiles in CODIS has increased significantly over recent years. This, in turn, increased the usefulness of DNA evidence in testing suspect DNA profiles against a well-populated database of existing offenders. This usefulness has increased awareness among law enforcement personnel of CODIS, which contributes to increased demand for DNA analysis, thereby contributing to the backlog. Additionally, when deciding whether to submit DNA evidence for analysis, law enforcement and prosecutors may consider jurors' expectations that DNA analysis is presented.[19]

Recent Legislation Requiring Sexual Assault Kit (SAK) Analysis

State legislation requiring SAK analysis has caused an increase in demand for DNA analysis. As of July 2018, we identified at least 25 states that have enacted laws requiring law enforcement to submit for testing SAKs that come into law enforcement possession.[20] Eleven of these states also required the submission for testing of previously untested SAKs. Twenty-one of these laws were passed in 2014 or later.

In addition to the factors that have contributed to increased demand, resource challenges and constraints on lab capacity are reported to have helped contribute to crime scene evidence backlogs. State and local labs generally receive appropriations from state or local governments and are subject to local funding priorities.

"free" for submitting law enforcement agencies. In this context, turnaround time may be a substitute for price—and thus when turnaround time decreases, it can be expected that quantity demanded from law enforcement will increase in response. Further, laboratory capacity may be a substitute for supply.

[19] The literature and stakeholders commonly refer to this as the "CSI effect," which is named after a television program that features the use of DNA in solving criminal cases.

[20] Legislation varies by state and may provide for the testing of SAKs or other evidence relating to sexual assault consistent with prioritization policies or other requirements, such as the sexual assault, be reported to law enforcement.

Federal grants can help, but even combined federal and jurisdictional funding may not increase lab capacity enough to keep up with increases in demand. Additionally, these labs report facing lengthy hiring and training processes for forensic analysts, and often lose staff to private or federal labs which may offer higher pay, further limiting lab capacity for completing analysis.

PRELIMINARY RESULTS SHOW THAT DOJ HAS NOT CLEARLY DEFINED AND DOCUMENTED CEBR GRANT PROGRAM GOALS

DOJ's NIJ has not defined CEBR program-wide goals in clear, specific, and measurable terms.[21] We identified statements in NIJ and CEBR program documentation that communicated program-wide goals, but the documentation did not consistently identify the same goals or cite the same number of goals. For example, a stated goal of improving the quality of DNA testing was included in only 2 of 4 NIJ documents we reviewed. In addition, NIJ officials verbally clarified that the CEBR program has two goals, (1) to increase laboratory capacity for DNA analysis, and (2) to reduce backlogs of DNA evidence awaiting analysis. These differences can be seen across goal statements outlined in various NIJ sources as shown in Table 1 below.

NIJ officials acknowledged that they do not have documentation that further defines the goals of the program in clear, specific, and measurable terms. These goals are specified as increasing laboratory capacity for DNA analysis and reducing backlogs of DNA evidence awaiting analysis. Officials provided an explanation as to what the goals mean.

[21] Program-wide goals are different than goals for individual grants. The CEBR grant program allows grantees to create their own goals and objectives within the framework of overall program-wide goals.

Table 1. DNA Capacity Enhancement and Backlog Reduction Program (CEBR) Goals Listed in National Institute of Justice (NIJ) Documentation and NIJ Verbal Clarifications

Source	**CEBR Goal(s)**
Goals as stated in NIJ documentation	
CEBR grant solicitation (fiscal year 2018)	To assist eligible States and units of local government to increase laboratory capacity and reduce the number of samples awaiting analysis in both the casework and database sections.[a]
NIJ website (2018)[b]	To assist eligible states and units of local government to: • Process, record, screen and analyze forensic DNA and/or DNA database samples. • Increase the capacity of public forensic DNA and DNA database laboratories to process more DNA samples, thereby helping to reduce the number of forensic DNA and DNA database samples awaiting analysis.
NIJ report (2018)[c]	To improve the quality of testing and increase the efficiency of evidence processing in forensic DNA laboratories.
NIJ report (2017)[d]	To help increase the throughput of evidence processing at the nation's forensic DNA laboratories and improve the quality of testing.
NIJ verbal clarification of goals	
NIJ verbal explanation (2018)	The CEBR program has two goals: 1. Increase laboratory capacity for DNA analysis. 2. Reduce backlogs of DNA evidence awaiting analysis.

Source: GAO analysis of NIJ documentation and interview with NIJ officials. | GAO-18-651T.

[a]"Sections" refers to areas within labs that perform different types of analysis; the "casework" section refers to the section that performs DNA analysis on evidence from crime scenes, victims, and suspects; the "database" section refers to the section that performs DNA analysis on samples collected from convicted offenders, arrestees, and others as authorized by law.

[b]DOJ, OJP, NIJ, DNA Capacity Enhancement and Backlog Reduction Program, https://www.nij.gov/topics/forensics/lab-operations/evidence-backlogs/Pages/backlog-reductionprogram.aspx. Accessed June 4, 2018.

[c]DOJ, OJP, NIJ, Report Forensic Science Fiscal Year 2017 Funding for DNA Analysis, Capacity Enhancement, and Other Forensic Activities. (Washington D.C.: April 2018).

[d]DOJ, OJP, NIJ, Report Forensic Science Fiscal Year 2016 Funding for DNA Analysis, Capacity Enhancement, and Other Forensic Activities. Washington D.C.: May 2017).

Specifically, officials stated that:

- Increasing lab capacity refers to increasing samples analyzed, reducing processing times, and increasing the number of DNA profiles uploaded into CODIS—all while either maintaining or increasing the quality of DNA analysis at labs.

- Reducing backlogs refers to reducing the number of backlogged requests awaiting analysis by more than the number of requests that become backlogged during the same timeframe. Officials stated that although they believe the goal of reducing the crime scene evidence backlog is unachievable in the foreseeable future, they have kept it as a program goal because each year it is included in the appropriation language that supports the program.

However, these clarifications and definitions are not available in CEBR documentation, which is an indication that NIJ may not be using clear, specific, and measurable goals to guide program development or assess progress.[22]

We continue to evaluate CEBR program goals and we are in the process of evaluating related CEBR performance measures as part of our ongoing work.

PRELIMINARY ANALYSIS SHOWS THAT OJP HAS ESTABLISHED CONTROLS FOR CONFLICTS OF INTEREST RELATED TO CEBR GRANTS, BUT HAS NOT FULLY ESTABLISHED CONTROLS RELATED TO LOBBYING

Our preliminary results show that OJP has controls to implement federal requirements associated with conflicts of interest and some controls

---

[22] In a 2013 report, we assessed the extent to which NIJ verified data on grant results submitted by grantees and measured outcomes of the DNA Backlog Reduction Program (now called the CEBR program). We found that NIJ did not have an approach to verify grantee performance data and that one of its performance measures included estimated rather than actual outcomes. As a result of recommendations we made in this report, which we have since closed as "implemented," NIJ now has a process in place to validate selected performance measures and replaced one of its performance measures to better reflect increasing laboratory capacity. GAO, *Justice Grant Programs: DOJ Could Improve Decision-Making Documentation and Better Assess Results of DNA Backlog Reduction Program*. GAO-13-605, (Washington, D.C: July 2013).

related to lobbying that apply to both OJP CEBR grant administrators as well as recipients of grant funding; however, OJP has not fully established all appropriate controls related to lobbying.

OJP Has Established Controls for Conflicts of Interest Related to CEBR Grants

We found that OJP has established controls to implement federal conflicts of interest requirements that apply to OJP employees administering CEBR grants and CEBR grantees. For example, federal law prohibits government employees from participating personally and substantially in particular government matters, such as the administration of federal grants, which could affect their financial interests.[23] We found that OJP has established an agency-wide ethics program and uses tools such as the DOJ Ethics Handbook and annual financial disclosure reports, among others, to help employees and their supervisors to determine whether they have potential conflicts of interest. See Table 2 below for a list of the federal conflicts of interest requirements we identified, as well as our preliminary assessment of related OJP controls to ensure that the requirements are met.[24]

---

[23] Criminal conflict of interest statutes governing OJP employees who administer CEBR grants are codified at 18 U.S.C. Chapter 11. Although these statutes cover a variety of topics related to conflicts of interest, our review focuses on the participation of OJP employees in government actions that may conflict with their personal financial interests, as specifically provided in 18 U.S.C. § 208. We focus on acts affecting personal financial interests because related regulatory requirements cover a broad range of issues directly applicable to OJP employees who administer CEBR grants. These regulatory requirements are set forth in selected subparts of 5 C.F.R. pt. 2635 "Standards of Ethical Conduct for Employees of the Executive Branch."

[24] Because OJP administers CEBR as a formula grant program in which applicants are entitled to a specific amount of federal funds if they meet certain requirements, some risks associated with conflicts of interest that affect competitive grants, such as peer reviewer conflicts of interest, do not apply to CEBR.

OJP Has Some Controls for Lobbying as They Apply to Recipients of CEBR Grant Funds, but Has Not Fully Established All Appropriate Controls

We found that OJP has established some controls related to lobbying but has not fully established controls needed to meet applicable requirements. Specifically, federal law sets forth several requirements related to lobbying "certification" and "disclosure."[25] Lobbying certification refers to agreeing not to use appropriated funds to lobby, and lobbying disclosure refers to disclosing lobbying activities with respect to the covered federal action paid for with nonappropriated funds. Federal regulation requires recipients of all federal awards over $100,000 to file certification documents and disclosure forms (if applicable) with the next tier above, and to forward those same forms from the tier below if they issue subawards for $100,000 or more.[26] In the case of CEBR grants, tiers include OJP, grantees, subgrantees, contractors under grantees and subgrantees, and subcontractors. Subawards include subgrants, contracts under grants or subgrants, and subcontracts.

We found that OJP had established controls to obtain lobbying certification documents and disclosure forms from grantees, but had not fully established controls to ensure grantees obtain these documents from tiers below them, see Table 3 below.

---

[25] 28 C.F.R. pt. 69, "New Restrictions on Lobbying," implements 31 U.S.C. § 1352, commonly referred to as the "Byrd Amendment." Lobbying in the context of this statute refers to paying any person for influencing or attempting to influence an officer or employee of any agency, a Member of Congress, an officer or employee of Congress, or an employee of a Member of Congress in connection with any covered Federal actions. See also, 2 C.F.R. pt. 200, Appendix II(I) which contains similar requirements.

[26] The required certification document is set forth in Appendix A to 28 C.F.R. pt. 69. The required disclosure form is set forth in appendix B to 28 C.F.R. pt. 69. Disclosure forms are only required if the recipient has used or plans to use nonappropriated funds to lobby with respect to the award. Additional exceptions to the prohibition and disclosure requirements apply.

Table 2. Preliminary Analysis of Office of Justice Programs (OJP) Controls to Ensure that OJP Employees and Grantees Meet Federal Requirements Related to Conflicts of Interest

Relevant statute or regulation	Select requirements for OJP employees and Capacity Enhancement and Backlog Reduction program grantees	Preliminary analysis of OJP controls to ensure that requirements are met
5 C.F.R. pt. 2638, "Executive Branch Ethics Program"	Agencies must carry out a government ethics education program. New employees must complete initial ethics training. Employees must complete annual ethics training.	•
5. C.F.R. pt. 2635, "Standards of Ethical Conduct for Employees of the Executive Branch"	Employees must not participate in matters likely to affect their financial interests or under circumstances that would cause a reasonable person to question their impartiality. Employees may not participate in particular matters that have a direct and predictable effect on the financial interests of persons with whom employees are seeking employment. Employees are not allowed to engage in outside employment or other outside activities that conflict with their official duties.	•
5 U.S.C. app. §§ 101-111, "Financial Disclosure Requirements of Federal Personnel" 5 C.F.R. pt. 2634, "Executive Branch Financial Disclosure, Qualified Trusts, and Certificates of Divestiture"	Senior-level employees must file public financial disclosure reports annually. Specified less senior employees must file confidential financial disclosure reports annually.	• •
2 C.F.R. pt. 200, "Uniform Administrative Requirements, Cost Principles, and Audit Requirements for Federal Awards"	Agencies must establish conflicts of interest policies for grants. Grantees must disclose in writing any conflicts of interest to the agency.	•

Legend: Control(s) fully established ●. Control(s) partially established ◖.

Source: GAO analysis of conflicts of interest requirements and OJP information. | GAO-18-651T.

Table 3. Preliminary analysis of Office of Justice Programs (OJP) Controls to Ensure that Grant Award Recipients Meet Federal Requirements Related to Lobbying

Relevant federal statute or regulation	Select requirements for Capacity Enhancement and Backlog Reduction program (CEBR) grantees and others	Preliminary Analysis of OJP controls to ensure that requirements are met
31 U.S.C. § 1352 28 C.F.R. pt. 69, "New Restrictions on Lobbying"	OJP is to collect certification documents and disclosure forms[a] from CEBR grantees.[b]	●
If grantees issue subawardsc over $100,000, OJP is to ensure grantees collect certification documents and disclosure forms from tiersd below them. OJP is to ensure that disclosure forms are forwarded from tier to tier until received by OJP.	If grantees issue subawards[c] over $100,000, OJP is to ensure grantees collect certification documents and disclosure forms from tiers[d] below them. OJP is to ensure that disclosure forms are forwarded from tier to tier until received by OJP.	◖

Legend: Control(s) fully established ●.

Control(s) partially established ◖.

Source: GAO analysis of 31 U.S.C. § 1353 and 28 C.F.R. pt. 69 as applicable to CEBR grantees and OJP information. | GAO-18-651T.

[a]Disclosure forms are only required by regulation if the award recipient or subrecipient used nonappropriated funds to lobby with respect to the CEBR grant. However, we found that OJP requires all CEBR grantees to submit a disclosure form when applying for a CEBR grant.

[b]This requirement is only applicable to federal awards over $100,000. However, all CEBR grants are for at least $150,000. Additional exceptions and requirements apply.

[c]In the case of CEBR grants, subawards include subgrants, contracts under grants or subgrants, and subcontracts.

[d]In the case of CEBR grants, tiers include OJP, grantees, subgrantees, contractors under grantees and subgrantees, and subcontractors; and subawards include subgrants, contracts under grants or subgrants, and subcontracts.

OJP has established mechanisms to ensure it obtains lobbying certification documents and disclosure forms from grantees. Specifically, according to OJP, it requires that grant applicants electronically agree to the certification document during the application process; if applicants do not agree to it, they cannot move on in the process. OJP also requires that applicants submit the lobbying disclosure form as part of the grant application process. Upon submission, a grant manager reviews the form for completeness and content and checks a box in an application review checklist.

However, OJP has only partially established a mechanism to ensure that, for subawards over $100,000 (1) CEBR grantees obtain certification documents and disclosure forms, as applicable, from tiers below them, and (2) disclosure forms are forwarded from tier to tier until received by OJP. Specifically, OJP requires grant applicants to agree to the certification document set forth in regulation.[27] This certification document, in turn, lists certification and disclosure requirements, and states that, "The undersigned shall require that the language of this certification be included in the award documents for all subawards at all tiers (including subgrants, contracts under grants and cooperative agreements, and subcontracts) and that all subrecipients shall certify and disclose accordingly." However, the certification document does not state in clear terms what the specific requirements of the regulation are or how they are to be carried out.

OJP attorneys responsible for overseeing their implementation were not aware of specific requirements in the regulation. For example, they were not aware that disclosure forms were required to be forwarded from tier to tier until received by OJP. Additionally, 3 of 4 CEBR grantees we spoke with were not aware of one or more of these requirements.

Lastly, we found that OJP does not provide guidance to grantees to ensure they understand the requirements nor does OJP follow-up with grantees to ensure they are implementing them. The statute requires that federal agencies "take such actions as are necessary to ensure that the [lobbying requirements] are vigorously implemented and enforced in [that] agency."[28] As part of our ongoing work, we will continue to monitor and assess OJP's compliance with statute and regulations related to grantee, subgrantee, and contractor lobbying disclosure requirements and make recommendations, as appropriate.

Chairman Grassley, Ranking Member Feinstein, and Members of the Committee, this concludes my prepared statement. I would be pleased to respond to any questions that you may have at this time.

---

[27] Appendix A to 28 C.F.R. pt. 69.

[28] 31 U.S.C. § 1352(f); *see also* 31 C.F.R. § 21.410.

In: DNA
Editor: Tomáš Koláček
ISBN: 978-1-53616-117-5

Chapter 6

SEXUAL ASSAULT: INFORMATION ON THE AVAILABILITY OF FORENSIC EXAMINERS*

Nicole Clowers

WHY GAO DID THIS STUDY

In 2016, about 323,000 individuals age 12 or older were reported victims of sexual assault, according to the Bureau of Justice Statistics. Studies have shown that exams performed by sexual assault forensic examiners—medical providers trained in collecting and preserving forensic evidence—may result in better physical and mental health care for victims, better evidence collection, and higher prosecution rates. Yet, concerns have been raised about the availability of such examiners. The Department of Justice administers grant programs that can be used by states and other eligible entities to train and fund examiners.

This statement summarizes GAO's findings from its March 2016 report (GAO-16-334) describing (1) what was known in 2016 about the

* This is an edited, reformatted and augmented version of the United States Government Accountability Office Testimony before the Subcommittee on Oversight and Investigations, Committee on Energy and Commerce, House of Representatives, Publication No. GAO-19-259T, dated December 12, 2018.

availability of sexual assault forensic examiners nationally and in selected states and (2) the challenges selected states faced in maintaining a supply of sexual assault forensic examiners. For that report, GAO reviewed literature on the availability of examiners and challenges training and retaining them. GAO also interviewed knowledgeable officials, including recipients of federal sexual assault examiner related grants and officials from sexual assault coalitions in six states (Colorado, Florida, Massachusetts, Nebraska, Oregon, and Wisconsin) selected to achieve variation in factors such as population and geographic location.

WHAT GAO FOUND

GAO's March 2016 report examining the availability of sexual assault forensic examiners found that only limited nationwide data existed on the availability of sexual assault forensic examiners—both the number of practicing examiners and health care facilities that had examiner programs. At the state level, GAO found that, in three of the six states it selected to review, grant administrators or officials from sexual assault coalitions were able to provide estimates of the number of practicing examiners and, in all six states, they were able to provide information on the estimated number of examiner program locations in their state. However, officials in all six selected states told GAO that the number of examiners available in their state did not meet the need for exams, especially in rural areas. For example, officials in Wisconsin explained that nearly half of all counties in the state did not have any sexual assault examiner programs available and officials in Nebraska told GAO that most counties in the state did not have examiner programs available. As a consequence, officials said victims may need to travel long distances to be examined by a trained examiner. In health care facilities where examiners were available, they were typically available in hospitals on an on-call basis, though the number available varied by facility and may not provide enough capacity to offer examiner coverage 24 hours, 7 days a week.

GAO's March 2016 report also found there were multiple challenges to maintaining a supply of examiners, according to its review of the literature and interviews with officials in the six selected states. These challenges include:

- *Limited availability of training*. Officials in five of the six selected states reported that the limited availability of classroom, clinical, and continuing education training opportunities is a challenge to maintaining a supply of trained examiners. For example, officials told us that there is a need for qualified instructors to run training sessions.
- *Weak stakeholder support for examiners.* Officials in five of the six selected states reported that obtaining support from stakeholders, such as hospitals, was a challenge. For example, hospitals may be reluctant to cover the costs of training examiners or pay for examiners to be on call.
- *Low examiner retention rates.* The above-mentioned and other challenges, including the emotional and physical demands on examiners, contribute to low examiner retention rates. Officials in one of the selected states estimated that while the state trained 540 examiners over a two-year period, only 42 of those examiners were still practicing in the state at the end of those 2 years.

Officials described a variety of strategies they have employed that can help address these challenges, such as implementing web-based training courses, clinical practice labs, mentorship programs, and multidisciplinary teams that respond to cases of sexual assault.

Chairman Harper, Ranking Member DeGette, and Members of the Subcommittee:

I am pleased to be here today to discuss our work on the availability of sexual assault forensic examiners. An estimated 323,450 individuals age 12 or older were victims of rape or other sexual assault in 2016, according to the most recently available data from the Bureau of Justice Statistics.[1] When victims of sexual assault receive a medical forensic examination, the exam may be provided by either a trained sexual assault forensic examiner—that is, a medical provider who has received specialized training in properly collecting and preserving forensic evidence—or a medical provider who has not received such specialized training. Studies have shown that exams performed by trained sexual assault forensic examiners may result in shortened exam time, higher quality health care delivered to victims, higher quality forensic evidence collection, as well as better collaboration with the legal system and higher prosecution rates. However, concerns have been raised about the availability of examiners to meet victims' needs for exams.

To help inform today's discussion, my testimony will focus on findings from our March 2016 report examining information on the training, funding, and availability of sexual assault forensic examiners.[2] In particular, this statement will address:

1. what was known about the availability of sexual assault forensic examiners nationally and in selected states as of 2016, and
2. the challenges selected states faced in maintaining a supply of sexual assault forensic examiners.

For our March 2016 report, we conducted a literature review to identify studies that measured the availability of sexual assault forensic examiners, examined challenges to training and retaining examiners, and strategies that could be used to address these challenges. We interviewed experts, recipients of federal grants to train sexual assault forensic examiners, and state sexual assault coalition officials in six selected states about data on the availability of examiners or examiner programs; the extent to which examiner availability meets the need for exams; challenges

---

[1] Data presented are from the Bureau of Justice Statistics' 2016 National Crime Victimization Survey. See R.E. Morgan and G. Kena, *Criminal Victimization, 2016*, NCJ 251150 (Washington, D.C.: Department of Justice, December 2017).

[2] See GAO, *Sexual Assault: Information on Training, Funding, and the Availability of Forensic Examiners*, GAO-16-334 (Washington, D.C.: March 18, 2016).

they experienced in training or retaining examiners; and strategies that could be used to overcome these challenges.[3] We also interviewed officials from the International Association of Forensic Nurses (IAFN) about these issues. Our March 2016 report includes a full description of our scope and methodology. We conducted the work on which this statement is based in accordance with generally accepted government auditing standards. Those standards require that we plan and perform the audit to obtain sufficient, appropriate evidence to provide a reasonable basis for our findings and conclusions based on our audit objectives. We believe that the evidence obtained provides a reasonable basis for our findings and conclusions based on our audit objectives.

BACKGROUND

Victims of sexual assault may receive a sexual assault forensic examination by a medical provider who may or may not be a trained sexual assault forensic examiner. Medical providers assess victims' clinical conditions; provide appropriate treatment and medical referrals; and, given consent by the victim, collect forensic evidence through a sexual assault forensic examination that may follow steps and use supplies from a sexual assault evidence collection kit. Under its protocol for sexual assault forensic examinations, the Department of Justice (DOJ) recommends that medical providers collect a range of physical evidence.[4]

---

[3] State sexual assault coalitions of rape crisis centers and other organizations provide direct support to members through funding, training and technical assistance, public awareness activities, and public policy advocacy. To select the six states (Colorado, Florida, Massachusetts, Nebraska, Oregon, and Wisconsin), we considered the number of grantees in each state that received funding from selected federal grant programs; whether states had unique policies or programs in place regarding the training of examiners; and state population size and geographic location. We sought to achieve variation in these characteristics when selecting the six states. In the six selected states, we interviewed a total of nine grantees that received federal funds in fiscal year 2014. Information from these interviews cannot be used to generalize beyond the six selected states.

[4] Department of Justice, *A National Protocol for Sexual Assault Medical Forensic Examinations: Adults/Adolescents*, NCJ 228119 (Washington, D.C.: U.S. Department of Justice, 2013).

In addition, sexual assault forensic exams typically include documenting biological and physical findings such as cuts or bruises and a victim's medical forensic history, such as the time and nature of the assault. Once the exam is complete, medical providers preserve the collected evidence, which may include packaging, labeling, and sealing evidence collection kits and storing kits in a secure location. Medical providers typically perform such exams only for acute cases of sexual assault, such as in cases where the assault occurred within the previous 72 to 96 hours, when the physical and biological evidence on a person's body or clothes is considered most viable.

DOJ, IAFN, and the American College of Emergency Physicians (ACEP) recommend that sexual assault forensic exams be performed by specially trained medical providers—known as sexual assault forensic examiners (examiners). These examiners include physicians, physician assistants, nurse practitioners, and other registered nurses who have been specially educated and have completed clinical requirements to perform sexual assault forensic exams. Sexual assault nurse examiners (SANE) —a particular type of sexual assault forensic examiner—are registered nurses, including nurse midwives and other advanced practice nurses, who have received specialized education and have fulfilled clinical requirements to perform sexual assault forensic exams. Examiner programs have been created in hospital or non-hospital settings whereby specially trained examiners are available to provide first-response care and exams to sexual assault victims. DOJ, IAFN, and some states have issued guidelines pertaining to the minimum level of training examiners should receive in order to properly collect and preserve evidence, identify victims' medical and emotional health care needs, and provide counseling and referrals for victims. These guidelines include recommendations of objectives and topics that training programs should cover.

DOJ administers several grant programs that aim to, among other things, improve response to and recovery from four broad categories of victimization—domestic violence, sexual assault, dating violence, and stalking. In our March 2016 report we describe three key grant programs administered by DOJ's Office on Violence Against Women that could be used by grant recipients—including states or other eligible entities—to fund or train sexual assault forensic examiners.[5]

NATIONWIDE DATA ON THE AVAILABILITY OF SEXUAL ASSAULT FORENSIC EXAMINERS ARE LIMITED; OFFICIALS IN SELECTED STATES REPORTED A NEED FOR ADDITIONAL EXAMINERS

In our March 2016 report examining the availability of sexual assault forensic examiners, we found that only limited nationwide data exist on the availability of sexual assault forensic examiners—that is, both the number of practicing examiners and health care facilities that have examiner programs. While IAFN reported that, as of September 2015, there were 1,182 nurses with an active IAFN SANE certification in the United States, such data do not represent all practicing examiners nationwide. For example, the data do not account for examiners who completed training through an IAFN or a state training program but never became certified or were certified through another entity, such as a state board of nursing. IAFN also collects data on examiner programs nationwide—that is, data on hospitals, clinics, and other sites where examiners practice. Such data provide an indication of the availability of examiners, but the data are also limited. While 703 examiner programs nationwide voluntarily reported to IAFN's examiner program database, as of September 2015, IAFN officials noted that the database is often not up to date; and some health care settings where sexual assault forensic exams are conducted, such as child advocacy centers, are not represented. In addition, data collected on staffing characteristics of examiner programs are often unavailable in the IAFN examiner program database. For example, only about one-third of the examiner programs reported on the number of examiners practicing in

[5] These grant programs were the Services-Training-Officers-Prosecutors Violence Against Women Formula Grant Program; Grants to Encourage Arrest Policies and Enforcement of Protection Orders Program; and the Rural Sexual Assault Domestic Violence, Dating Violence, and Stalking Assistance Program. See GAO-16-334 for additional information on the use of these grant programs to train and fund sexual assault forensic examiners.

their program, and about one-third reported on whether examiners were available on-site versus on-call.

In three of the six selected states we reviewed in our March 2016 report, grant administrators or officials from sexual assault coalitions were able to provide estimates of the number of practicing examiners, and, in all six states, they were able to provide information on the estimated number of examiner program locations in their state. Of states that reported, the number of practicing examiners and examiner programs varied by state. (See table 1.) However, such data may also present an incomplete picture of the availability of examiners. For example, only one of the six selected states has a system in place to formally track the number and location of examiners. Instead, officials generally reported on the estimated number of examiners or examiner locations that were part of a statewide examiner program or were identified through an ad hoc data collection effort.

Table 1. Estimated Number of Practicing Sexual Assault Forensic Examiners and Examiner Programs in Selected States, As of January 2016

State	Estimated number of practicing examiners	Estimated number of examiner program locations
Colorado	Not Available	23
Florida	Not Available	15[a]
Massachusetts	141	29
Nebraska[b]	61	7
Oregon	140	12[c]
Wisconsin	Not Available	18

Source: GAO analysis of state data. | GAO-19-259T

[a] The reported number of examiner programs is limited to those located in certified rape crisis centers in Florida. It does not include examiner programs that are located in other facilities, such as hospitals.

[b] Data presented for Nebraska does not account for examiners who may be located in child advocacy centers. There are seven child advocacy centers in Nebraska that provide sexual assault forensic exams, including to adult victims.

[c] One of the 12 examiner programs is a mobile examiner program that serves five counties in Oregon.

Although data are limited, grant administrators and sexual assault coalition officials in all six selected states nevertheless told us that the number of examiners available does not meet the need for exams within their states. For example, coalition officials in Wisconsin told us that nearly half of all counties in the state do not have any examiner programs available, and coalition officials in Nebraska told us that most counties in the state do not have examiner programs available. In addition, in four of the six selected states—Colorado, Florida, Nebraska, and Wisconsin—state grant administrators and coalition officials told us that few or some health care facilities in their state have examiners available. As a consequence, officials said victims may need to travel long distances to be examined by a trained examiner or be examined by a medical professional without specialized training. While in the other two selected states—Massachusetts and Oregon—state grant administrators and coalition officials stated that some or most facilities have examiners available, they noted that there is still a need for additional capacity to reduce the burden on those examiners who are available, or to make examiners available in a number of areas where examiners are currently unavailable.

In health care facilities where examiners are available, they are typically available through hospitals on an on-call basis, according to literature we reviewed as well as all grant administrators and coalition officials we interviewed for our report. In addition, among facilities that have examiners available, the number of examiners available varies and may not provide enough capacity for facilities to offer examiner coverage 24 hours, 7 days a week, according to state grant administrators and coalition officials we interviewed. Nebraska coalition officials, for example, told us that while one hospital in Omaha has a team of 26 examiners available, other facilities in the state may have as few as three examiners available. Further, officials from Florida and Colorado told us that there are few facilities in their states able to offer full coverage with examiners available 24 hours, 7 days a week.

SELECTED STATES FACED CHALLENGES TRAINING EXAMINERS, MAINTAINING STAKEHOLDER SUPPORT, AND RETAINING EXAMINERS

In our March 2016 report, we found that maintaining a supply of trained examiners that meets communities' needs for exams is challenging for multiple reasons, and that state officials have employed a variety of strategies to address these challenges, as described below.

Limited Availability of Training

Officials in five of the six selected states told us that the limited availability of classroom, clinical, or continuing education training is a barrier to maintaining a supply of trained examiners. Regarding classroom training, some officials told us that training may only be offered once per year in their states. Additionally, officials from both Florida and IAFN told us that there is a need for qualified instructors to run training sessions. Experts and officials from Colorado, Nebraska, and Oregon also told us that medical professionals in rural areas may have difficulty completing the clinical training necessary to become an examiner. Obtaining clinical experience, such as performing exams under the supervision of a trained examiner, is a particular challenge in rural areas where hospitals may treat only a few sexual assault cases per year. One official in Nebraska told us that trained examiners in rural areas might not feel competent to perform exams due to the low number of cases they treat. A lack of continuing education opportunities may also pose a challenge for examiners in maintaining the skills necessary to perform exams. For example, the National Sexual Violence Resource Center (NSVRC) reported that—based on common challenges identified through a survey of, and group discussions among, examiner program coordinators—maintaining competency may be difficult for nurses in rural areas due to a low volume of patients presenting in need of exams and limited access to ongoing and advanced training.[6]

Officials told us they have been able to increase the availability of examiner training through alternative training methods such as web-based training courses and simulated clinical training. For example, officials in Colorado told us their state's web-based examiner training program has made training less expensive and has increased examiner recruitment. Officials in Wisconsin told us they developed a clinical training lab that allows examiners to gain hands-on experience by performing elements of exams on experienced teaching assistants hired for the purpose of training new examiners. Further, in 2014, a DOJ funded evaluation of examiner training programs found that a web-based training course may help increase the availability of trained examiners; the study also found that implementing web-based training had benefits such as decreasing the costs associated with attending in-person training, expanding training opportunities to remote areas, and allowing examiners to be trained by national experts.[7]

Lack of Technical Assistance and Other Supportive Resources

Officials in four of the six selected states told us that the limited availability of technical assistance and other supportive resources for examiners poses a challenge to maintaining a supply of trained examiners. For example, officials in Florida, Nebraska, Oregon, and Wisconsin explained that, in general, there is a lack of mentorship opportunities and leadership within the examiner community. Officials also noted that the sustainability of examiner programs may be threatened by a lack of internal capacity, such as not having a full-time, paid examiner program coordinator available. Further, in its survey of and group discussions with examiner

---

[6] National Sexual Violence Resource Center, First National SANE Coordinator Symposium: Final Report and Recommendations (Portland, Oregon, 2009).

[7] D. Patterson, S. Resko, J. Pierce-Weeks, and R. Campbell, Delivery and Evaluation of Sexual Assault Forensic (SAFE) Training Programs, Doc. No. 247081 (Washington, D.C.: Department of Justice, June 2014).

program coordinators, NSVRC found that examiners and examiner programs needed technical assistance and support in the following areas: aspects of performing exams, training, leadership development and policy issues, and examiner program sustainability.[8]

Officials we spoke to told us about strategies that can be used to increase support for examiners and examiner programs, such as offering web-based technical assistance. For example, officials in Massachusetts told us that, through their National Sexual Assault TeleNursing Center, trained SANEs provide remote clinical guidance to two hospitals in the state that do not have trained examiners available.[9] In addition, officials from Colorado told us an examiner program coordinator in an urban hospital in the state provides volunteer on-call technical assistance and clinical guidance to examiners in rural parts of the state, where those resources are not otherwise available. Further, one study we reviewed found several states were engaged in promising practices to increase support for examiners, such as implementing state-wide mentorship programs, developing regional examiner list-serves and online discussion boards, creating formal leadership positions within the examiner community, and requiring examiner program evaluations.[10]

Weak Stakeholder Support for Examiners

Officials in five of the six selected states told us that limited stakeholder support for examiners and examiner programs, such as from hospitals and law enforcement, is a challenge to maintaining a supply of trained examiners. Some officials told us that hospitals may be reluctant to support examiners and examiner programs due to a low number of sexual assault cases treated each year. One official told us that hospitals may be reluctant to send nurses to examiner training, as it takes away from their

[8] National Sexual Violence Resource Center, *First National SANE Coordinator Symposium*.

[9] The National Sexual Assault TeleNursing Center is funded by the DOJ Office of Justice Programs, Office for Victims of Crime, and is aimed at providing live access to expert medical forensic examiners via telemedicine.

[10] National Sexual Violence Resource Center, *First National SANE Coordinator Symposium*.

regular shift availability. Additionally, some hospitals do not pay examiners to be on call. Officials in three states told us that hospitals typically either do not pay examiners to be on call or pay on-call examiners significantly less than other on-call medical professionals.

Apart from hospital support, officials in Colorado and Oregon explained there is a need for more multidisciplinary support for examiners, such as increased law enforcement, prosecutor, and first-responder understanding of examiners' role. The literature we reviewed also shows that ambiguity around the role of the examiner in responding to sexual assault may be a source of conflict between examiners and other professionals.[11] For example, examiners were found to have experienced instances where victim advocates or law enforcement questioned examiners' medical decisions, speed of evidence collection, or asked examiners to comment on the credibility of a victim's case. One nationally representative survey of examiner programs found that examiner program coordinators felt ongoing education of community stakeholders on sexual assault and examiner programs was needed due to the high turnover in staff at relevant community institutions and agencies, such as law enforcement officers, victim advocates, and prosecutors.[12]

Through our interviews with officials, we learned of strategies selected states have used to increase or mitigate limited stakeholder support for examiners and examiner programs. For example, officials in Colorado, Florida, Nebraska, Oregon, and Wisconsin told us that sexual assault response teams have been developed in their states to help community stakeholders to understand examiners' role and better coordinate to meet the medical and legal needs of sexual assault victims.

---

[11] See, for example, R. Campbell, M. Greeson, and D. Patterson, "Defining the Boundaries: How Sexual Assault Nurse Examiners (SANEs) Balance Patient Care and Law Enforcement Collaboration," *Journal of Forensic Nursing* vol. 7, no. 1 (2011); J. Cole and T.K. Logan, "Negotiating the Challenges of Multidisciplinary Responses to Sexual Assault Victims: Sexual Assault Nurse Examiner and Victim Advocacy Programs," *Research in Nursing & Health* vol. 31, no. 1 (2008); and S. L. Maier, "Sexual Assault Nurse Examiners' Perceptions of Their Relationship with Doctors, Rape Victim Advocates, Police, and Prosecutors," *Journal of Interpersonal Violence* vol. 27, no. 7 (2012).

[12] T. K. Logan, J. Cole, and A. Capillo, "Sexual Assault Nurse Examiner Program Characteristics, Barriers, and Lessons Learned," Journal of Forensic Nursing vol. 3, no. 1 (2007).

Low Examiner Retention Rates

Officials in four of the six selected states told us that low examiner retention rates can be an impediment to maintaining a supply of trained examiners. In addition to the challenges of limited training opportunities, technical assistance and other supportive resources, and stakeholder support for examiners, the physically and emotionally demanding nature of examiner work contributes to low examiner retention rates. Further, studies have indicated that dissatisfaction with compensation, long work hours, and lack of support, among other things, may contribute to examiner burnout. Examiners typically work on call in addition to their full time jobs as, for example, emergency department nurses. Officials in Florida told us that examiners may be on call for 6-hour, 12-hour, or even 24-hour shifts. Further, one survey of examiner programs in Maryland found that examiners were required to be on call for an average of 159 hours per month.[13] Wisconsin officials estimated that, although 540 SANEs were trained over a 2-year period, only 42 (less than 8 percent) were still practicing in the state at the end of those 2 years.

Chairman Harper, Ranking Member DeGette, and Members of the Subcommittee, this concludes my prepared statement. I would be pleased to respond to any questions that you may have at this time.

---

[13] Maryland Coalition Against Sexual Assault, The State of the State: Sexual Assault Forensic Examiner (SAFE) Programs in Maryland (Arnold, Md.: 2012).

INDEX

A

B

C

D

E

F

M

N

O

P

Q

R

S

T

U

V